Areca catechu and Its Byproducts

Narendra Reddy · Pornanong Aramwit

Areca catechu and Its Byproducts

Springer

Narendra Reddy
Center for Incubation Innovation Research and Consultancy
Jyothy Institute of Technology
Bengaluru, Karnataka, India

Pornanong Aramwit
Faculty of Pharmaceutical Sciences
Chulalongkorn University
Bangkok, Thailand

ISBN 978-3-032-22133-9 ISBN 978-3-032-22134-6 (eBook)
https://doi.org/10.1007/978-3-032-22134-6

This Springer imprint is published by the registered company Springer Nature Switzerland AG
The registered company address is: Gewerbestrasse 11, 6330 Cham, Switzerland

Foreword

For centuries, nature has generously fulfilled the needs and, often, even the greed of humanity. Yet, in the brief span of about a hundred years, relentless and unregulated exploitation of natural resources, environment, and ecosystems has irrevocably altered the planet and the environment in which we live. The consequences of such changes, many perceived and many unexpected, are now manifesting as catastrophic impacts on climate and living conditions, directly affecting human and animal life. Climate change, resource depletion, socio-economic disruptions, and even political challenges pose a significant threat to sustaining basic human needs. Ensuring adequate food, energy, materials, and essential resources for future generations is likely to be one of humanity's biggest challenges.

Agriculture has always been and continues to be the foundation of human existence. Today, unfortunately, the shrinking availability of land, water, and other natural resources, unpredictable and extreme weather events, rising input costs, and other factors are making farming increasingly difficult. These factors also affect crop yield, quality, and profitability. In addition to the main produce, all agricultural crops generate substantial amounts of residues, byproducts, and coproducts both on-field and during post-harvest processing and use. It is estimated that 3.5-5 billion tons of dry agricultural residues are generated globally every year. Yet, less than 30% of these residues are utilized, and most of it is either buried or burnt, leading to environmental pollution.

Crop residues are inevitably generated and are renewable, sustainable, and low-cost resources. They contain carbohydrates, proteins, lignin, and several other valuable polymers, and discarding them means abandoning the land, water, energy, and human efforts invested in producing them. In contrast, utilizing crop residues can provide an environmentally friendly, sustainable, and renewable source for polymers and materials, thereby reducing dependence on non-renewable fossil fuel-based resources. Recognizing this potential, several countries and institutions have embraced concepts such as circular economy, no-agricultural waste, and waste valorization, which are increasingly been seen as both essential and mandatory.

Areca catechu is an important commercial crop, particularly in tropical and subtropical climates, with annual global areca nut production of about 2.7 million tons. Although cultivated in a few countries, the primary produce from the plant, the areca nut, is consumed globally and known for both its beneficial and harmful effects. Consumption of areca nut has been associated with oral submucous fibrosis, oral cancer, and heart and metabolic disorders. At the same time, beneficial effects include antioxidant, antiageing, antidiabetic, anti-inflammatory properties, and others. In addition to the nuts, the areca plant generates several valuable byproducts, including husks, leaf sheath, leaf, and even the tree trunk. Areca husks and sheaths have been extensively studied for their potential application in fiber, energy, environmental remediation, packaging, composites, medical uses, and more.

This book provides a comprehensive and integrated understanding of the chemical and physical properties of the areca nuts and constituents of the nuts. Additionally, properties of areca husks, leaf sheath, leaves, and tree trunk generated as byproducts and their diverse applications have been thoroughly discussed. This book, perhaps the first, provides an understanding of both the beneficial and adverse effects of areca nut consumption. More importantly, it elucidates the potential and possibilities of utilizing the byproducts and coproducts generated during the cultivation and processing of areca catechu. We hope that this book will contribute toward the global efforts in making agriculture more sustainable and promote the responsible use of natural resources, particularly crop residues, and inspire future research in this direction.

Competing Interests The authors have no competing interests to declare that are relevant to the content of this manuscript.

Contents

Chapter 1
Areca Catechu Production, Processing, and Uses

1.1 Need for Sustainable and Renewable Biomass

Economic, environmental, technological, social, and even political concerns are making it inevitable for humankind to transition from a non-renewable toward sustainable, green, and environmentally benign resources for food, fiber, energy, materials, and other needs. Such efforts are predominantly based on the shift from conventional linear fossil-based economy into biobased and circular economy, where minimization of waste and maximization of natural resources is given priority. Several countries have already adopted the circular economy concepts, and many more are in that process (Aragones 2024; Tsakalova 2024). One of the major concepts in a biobased economy is the valorization of crop residues and other biobased resources (biomass) (Jaroenkietkajorn 2024), which not only contributes to the bioeconomy but will assist in meeting many of the sustainable development goals proposed by the United Nations and adopted by most countries (Holland 2024; Lokesh 2018). Valorization of crop residues will also provide a solution to address the environmental and health issues related to their disposal.

Fibers and fibrous materials are high-value commodity products that not only account for major share of resource utilization but are also significant contributors to greenhouse gas emissions, water and soil pollution. Currently, about 130 million tons of fibers are produced every year with synthetic fibers accounting for nearly 58% of the market share, in which polyester forms the most prominent fiber. Comparatively, world natural fiber production was at 31 million tons in 2024, with cotton being the most common natural fiber and accounted for about 25 million tons (20% global fiber share) (https://textileexchange.org/news/textile-exchange-releases-2024-materials-market-report/). Common data on natural fibers and their production are given in Table 1.1, and the prominent sources, classification, and major producers of natural fibers are shown in Fig. 1.1. Concerningly, it is reported that the production of natural fibers has not increased since 2020 due to stagnated yields, increasing labor and other input costs, constraints on natural resources required to grow the

N. Reddy and P. Aramwit, *Areca catechu and its byproducts*,
https://doi.org/10.1007/978-3-032-22134-6_1

fiber crops and also due to environmental restrictions on processing of natural fibers into textiles and other products (https://marketdirectory.messefrankfurt.com/images/original/document_downloads/10000008202401/357700/1708772178590_1281153072.pdf). These limitations on producing and processing natural fibers are expected to increase further.

Hence, several efforts are being made to promote sustainable and environmentally friendly practices and to identify and develop new alternative sources that can support or substitute the natural fibers currently in the market. One of the most economically and environmentally feasible approaches to develop new natural fibers is by utilizing crop residues, agricultural byproducts and coproducts, and non-crop biomass as a source for fibers. These resources are available as byproducts and coproducts in large quantities at low cost and are also less resource-intensive to grow compared to fiber crops. Further, crop residues are lignocellulosic and are inevitably generated but have limited uses and are generally burnt or buried. It has been estimated that 3.8×10^9 t/a of crop residues are generated globally every year (Adnane 2024; Zhao 2024). In addition, when agriculture and forestry residues are combined, the theoretical potential of biomass available is estimated to be approximately 120 EJ yr^{-1} today and expected to increase to 140–170 EJ yr^{-1} by 2100. However, about 40% of this biomass is

Table 1.1 Some information on production, price, and profits related to natural fibers in 2023 (Natural Fiber Market Size & Share | Industry Report 2030)

Natural fiber	Average price received by farmers ($/kg)	Value of production, $ billion	Producing households, millions	Average earnings per household, $ thousands
Abaca	1.29	0.08	0.1	0.5
Agave fibers	0.99	0.04	0.0	1.0
Coir, without pith	0.21	0.24	0.6	0.4
Cotton lint	1.78	43.87	24.0	1.8
Other fiber crops	0.32	0.20	0.6	0.3
Flax, long fiber	1.09	0.19	0.0	19.0
True hemp, raw or retted	2.17	0.60	0.3	2.1
Jute, kenaf, and allied fibers	0.71	2.05	10.0	0.2
Kapok fiber	3.56	0.27	0.1	3.4
Ramie, raw or retted	1.13	0.01	0.0	1.1
Sisal, henequen, etc.	1.73	0.48	0.1	8.6
Silk, raw	7.75	0.70	2.0	0.3
Wool, clean	6.70	7.05	5.0	1.4
Other animal fibers	13.40	0.36	0.1	4.2

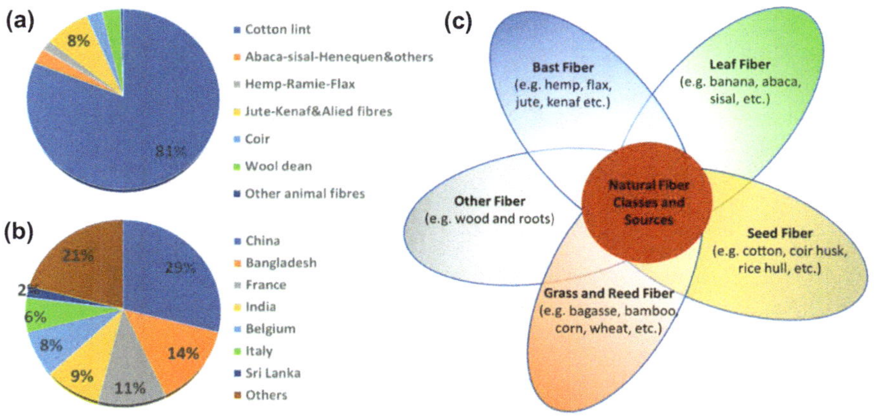

Fig. 1.1 Share of the different natural fibers (**a**), major producers (**b**), and common classification (**c**) of natural cellulose fibers (Jayaprakash, 2022). Reproduced with permission through open access publishing

desirable to remain on or in the ground for ecological reasons, and another 20% is subject to other uses, suggesting that at least approximately 50 EJ yr^{-1} in 2100 would be available at costs <10$\$_{2005}$ GJ^{-1} (Daioglou 2016). Hence, utilizing crop residues and other biomass is probably the most prudent approach to substitute for the production of natural fibers.

Many studies have been done to develop natural fibers from renewable and sustainable resources (El Bourakadi 2024). For instance, coir from coconut husks, fibers from stems of the banana plant are being converted into fibers for textiles and various other fibrous products, and are currently sold on the market. Similarly, sisal, bamboo, and other renewable resources have been used to produce natural fibers for commodity applications. Although these alternative natural fibers are available on the market and are being valorized into different products, they have either limited availability, high cost, or do not have the properties required to substitute cotton, flax, and other common natural fibers (Seki 2022). For example, coir fibers are available at less than $0.21 per kilogram compared to $1.78 for cotton, but their coarse nature and high lignin content make it difficult to be processed for normal textile applications. Nevertheless, coir has unique features and is extensively used for geotextiles, furnishings, and other applications. Several studies have also reported the feasibility of producing natural cellulose fibers from crop residues, including corn stover (Wubneh 2022; Reddy 2005, 2006), wheat straw (Reddy 2007), soybean straw (Reddy 2009), and the residues of other major food crops. Similarly, biomass such as switch grass and sabai grass, which are easy to grow, have also shown potential to provide fibers with properties similar to that of cotton and flax (Guna 2022; Reddy 2007). Hence, crop residues and biomass offer a sustainable, renewable, and low-cost resource for materials and products that currently rely on finite petroleum resources.

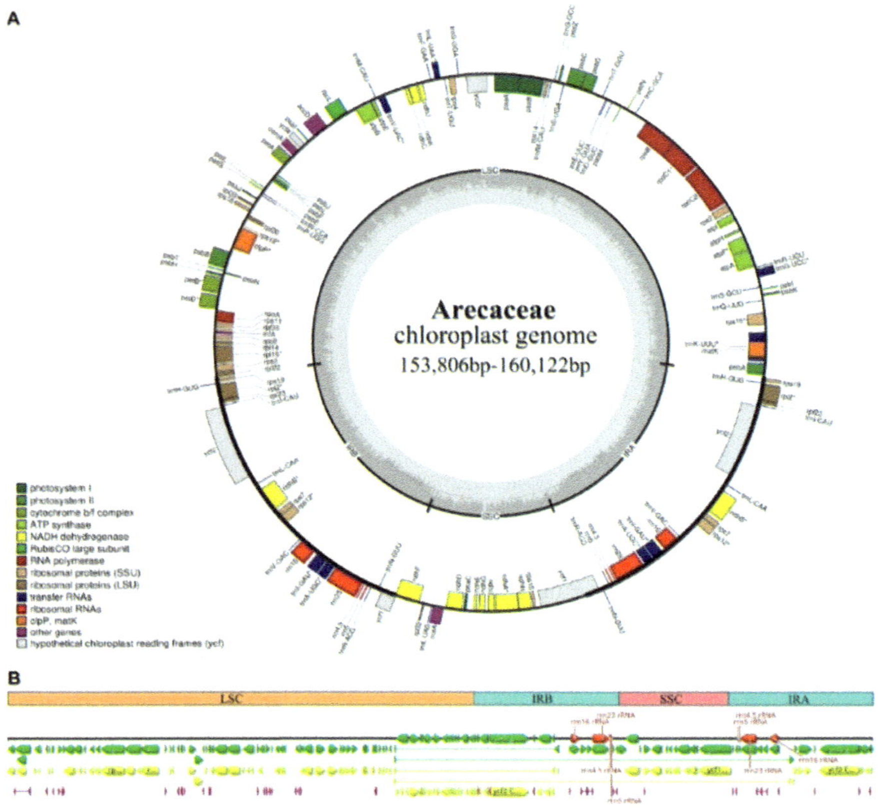

Fig. 1.2 Chloroplast genome map of Arecaceae. **a** Circular form. Genes inside the outer circles are transcribed clockwise, and those outside the circles are transcribed counterclockwise. The gray variation area in the inner circle indicates the GC content of the chloroplast genome. Different colors indicate different functional genes. **b** Linear form. Different colors indicate different functional genes, green indicates Genes, yellow indicates Protein Coding Genes, red indicates rRNA genes, and purple indicates tRNA genes (Chen 2022). Reproduced with permission through open access publishing

1.2 Areca Catechu Production and Availability

Areca (*Areca catechu* L. (Arecaceae)) is a plant widely found in South and Southeast Asia under tropical and sub-tropical conditions. Arecaceae belongs to the palm family Arecales and contains about 181 genera and about 2600 species, which can be further divided into five sub-families and 28 tribes. The arecaceae chloroplast genome is estimated to be between 153,860 and 160,122 base pairs (Fig. 1.2). Areca is considered a major commodity crop in many countries (Table 1.2), with India, Myanmar, Bangladesh, and China being the major producers. Areca trees grow up to 20–30 feet and typically have a life span of about 60 years. However, areca trees

Table 1.2 Major arecanut-producing countries and their annual production (Salehi 2020)

Country	Tons, 1000
India	996.07
Bangladesh	348.11
Myanmar	145.17
Indonesia	152.06
China	99.99
Sri Lanka	44.06
Thailand	38.14
Nepal	14.23
Bhutan	9.86
Malaysia	0.31
Kenya	0.11
Maldives	0.02

show considerable morphological differences depending on the sub-family, tribes, and location (Chen 2022) (Fig. 1.3).

Areca plants are mainly composed of the trunk, leaves, and leaf sheath, and the fruits contain the nuts, which are the primary commodity products. Areca nuts (Fig. 1.4) are typically harvested once a year, and the yield of areca nuts varies between 416 and as high as 8250 kg/ha (10–15 nuts/tree/year) depending on the species, age of the plant, climatic conditions, and other parameters. In terms of structure, arecanuts are enclosed in a protective fibrous layer typically called the husk. The fruit (Fig. 1.4) comprises of the nut and husks and weighs between 15 and 23 g. The husks, which form about 60–80% of the fresh fruit by weight, are removed (peeled or dehulled) and generally disposed of as waste.

1.3 Processing of Areca Nuts

Areca nuts are processed through several steps (Fig. 1.5) before they can be considered suitable for consumption (Guo 2024). The first step is the dehusking of the nuts, which is done either manually (Fig. 1.6) or using semi-automatic and automatic machines. Freshly peeled nuts are brownish in color and are juicy and highly fibrous, whereas the husks are green or golden yellow in color but are also highly fibrous (Fig. 1.7). These nuts are treated (usually steamed (Fig. 1.8)) to remove tannins, increase life, and also obtain desired taste and texture. After treatment, the nuts turn reddish brown and are ready to be stored for long-term use.

Fig. 1.3 Typical morphological characteristics of tribes of the subfamilies in Arecacea. Coryphoideae: (**A**) *Copernicia alba*; (**B, L, T**) *Phoenix roebelenii*; (**C, U**) *Sabal minor*; (**D**) *Borassus flabellifer*; (**E, M, V**) *Caryota mitis*; (**F, N, W**) *Chuniophoenix hainanensis*; (**K**) *Trachycarpus fortunei*; (**S**) *Lanonia dasyantha*; (**X**) *Bismarckia nobilis*; Arecoideae: (**G**) *Areca triandra*; (**H, Q, Z**) *Elaeis guineensis*; (**I**) *Hyophorbe lagenicaulis*; (**O**) *Wodyetia bifurcata*; (**P, Y**) *Chamaedorea elegans*; (**a**) *Ptychosperma macarthurii*; Calamoideae: (**J, R**) *Salacca zalacca*; (**b**) *Calamus rhabdocladus* (Chen 2022). Reproduced with permission through open access publishing

Fig. 1.4 Areca nuts harvested from the tree

1.4 Uses of Areca Nuts

Areca nuts are mainly used as a complement with beetle leaves and consumed orally in various forms and different trade names (Tables 1.3 and 1.4) (Mehrtash 2017). Areca nuts are composed of alkanoids, flavonoids, tannins, triterpenes, and steroids. The seeds are also notorious for the presence of arecoline and other compounds, which are carcinogenic, and hence areca nut use has been considered a major contributor to oral and other cancers in frequent and chronic users (Rumgay 2024; Garg 2024). About one-third of global lip and oral cavity cancer (LOC) are attributed to smokeless tobacco or areca nut consumption. A strong co-relation was also found between occurrence of cancer and areca nut growing regions and also areca nut importing regions (Cao 2025). In addition, areca nuts (specifically arecoline) are known for their pharmacological properties, including antiinflammatory and analgesic effects, antiallergic, antioxidant, antiparasitic, antibacterial, and antifungal effects, and hence extensively used for various medical applications (Peng 2015).

1.5 Byproducts from Areca Processing

The outer shell covering the nuts is called the husk and is the primary byproduct generated from areca processing. Husks are lignocellulosic in nature but with considerably high lignin content between 40 and 45%. To date, there are no major uses of areca husks, and most of them are disposed of by burning or burying (Fig. 1.9).

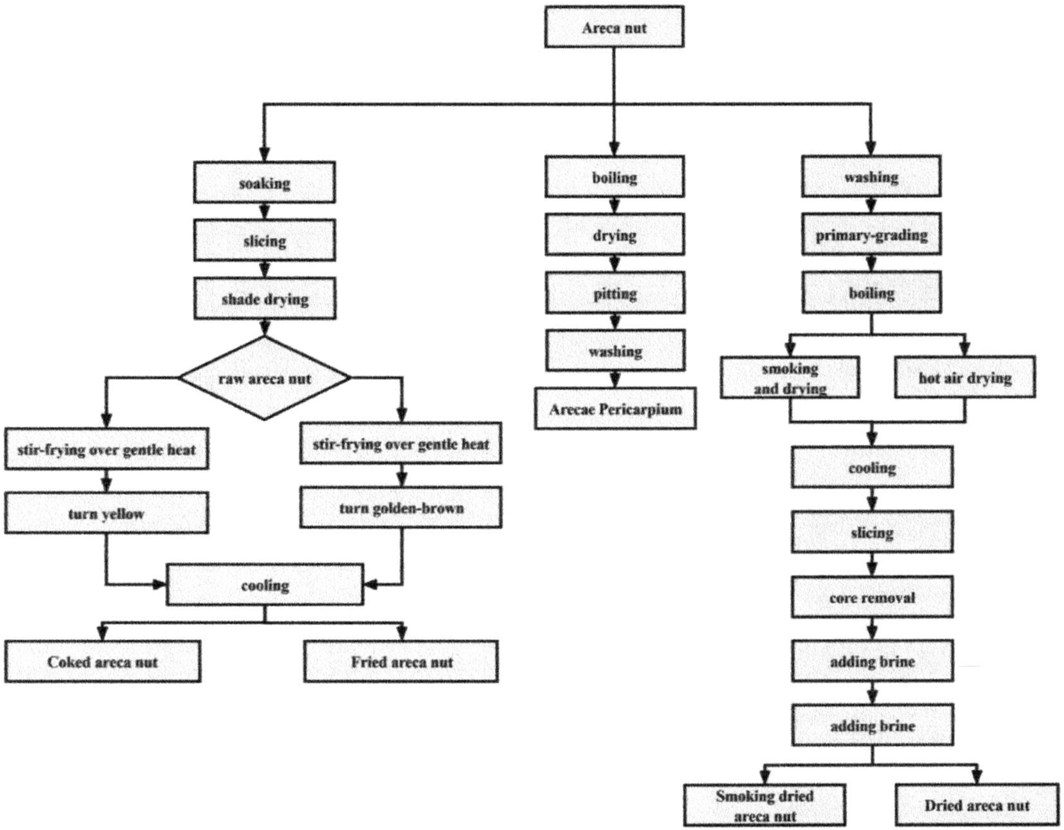

Fig. 1.5 Various approaches are used to process areca nuts for food and medical applications (Guo 2024). Reproduced with permission through open access publishing

Fig. 1.6 Manual (left) and mechanical (right) dehusking of areca nuts. Photo courtesy of Spoorthi B.R.

Fig. 1.7 Image of freshly peeled nuts (left) and separated husks (right). Photo courtesy of Spoorthi B.R.

Fig. 1.8 Steaming of areca nuts (left) and treated nuts ready for use (right). Photo courtesy of Spoorthi B.R.

Table 1.3 Some of the common betel quid, areca nut products, their common forms, and countries of use (Mehrtash 2017)

Form or product	Countries of use	Product definitions (form and type of tobacco and added ingredients)
Areca nut husk without tobacco	Mainland China	A custom-made or handmade product that consists of the husk of the Piper betle plant mixed with other ingredients (e.g., dried grapes)
Betel quid with tobacco (e.g., paan)	Bangladesh, Cambodia, China, Indonesia, India, Federated States of Micronesia, Laos, Malaysia, Maldives, Myanmar, Papua New Guinea, Nepal, Pakistan, Palau, Singapore, Sri Lanka, Taiwan, Thailand, United Arab Emirates, UK, Vietnam	Commercial or vendor-prepared packaged product, or handmade or home prepared by the user with tobacco; contains areca nut, slaked lime, betel leaf, and often catechu; other ingredients are added that differ regionally, such as cardamom, saffron, cloves, aniseed, turmeric, mustard, sweeteners, rosewater, aniseed, mint, or other spices
Betel quid (without tobacco)	Taiwan, Papua New Guinea, Solomon Islands	A custom-made or handmade product that is a combination of betel leaf, areca nut, or fruit of a pepper plant (Piper betle), and powdered lime; other ingredients and spices can be added to enhance flavoring; often prepared by wrapping the areca nut, and part of the betel pepper vine with or without the other ingredients in a Piper betel leaf
Gutkha or gutka	Bangladesh, India, Myanmar, Nepal, Pakistan, Sri Lanka	Commercially manufactured and packaged form of a tobacco product containing crushed areca nut and catechu that are mixed together with various flavorings and sometimes sweeteners
Khaini, khoini	Bangladesh, Bhutan, India, Nepal	Commercially prepared or custom-made product containing powdered tobacco mixed with slaked lime paste, and sometimes areca nut
Mainpuri	Uttar Pradesh, India	Handmade or custom-made products containing tobacco, areca nut, camphor, and cloves
Mawa, kharra	Gujarat, Maharashtra, India	Handmade or custom-made product or locally produced and wrapped in cellophane, containing small pieces of areca nut (95%) that are mixed with tobacco flakes and slaked lime, then the mixture is rubbed together to combine

(continued)

Table 1.3 (continued)

Form or product	Countries of use	Product definitions (form and type of tobacco and added ingredients)
Naswar (niswar)	Afghanistan, India, Pakistan	Custom-made form of tobacco that is a combination of ingredients, including slaked lime, ash, oil, sometimes flavorings (most commonly menthol and cardamom), and a coloring agent (such as indigo)
Zaarda or zarda	Bangladesh, Bhutan, India, Myanmar, Nepal, the UK, Yemen	Commercially produced product containing tobacco, lime, spices, vegetable dyes, and sometimes areca nut; processed with broken tobacco leaves boiled with lime and spices; the mixture is dried and colored with vegetable dyes, then sometimes mixed with areca nut

Reproduced with permission from Elsevier

Although areca nuts are the major produce, areca trees typically shed 5–6 leaves per year. Areca leaves have a unique sheath that attaches to the branch. The sheath forms about 10–20% of the leaf by weight and has a distinct structure and properties. Areca leaf sheaths are thick, impervious to water, and are lignocellulosic in nature. Although a majority of the sheaths are discarded as waste, a considerable number of sheaths are converted into packaging materials (Fig. 1.10). Apart from the sheath, most of the leaves are disposed of as waste or utilized as fertilizer.

Areca cultivation has social and economic prominence wherever it is grown, providing local employment, maintaining ecological balance, and sustaining livelihoods. However, the economic and social prominence of areca crops, areca cultivation, and processing are facing considerable restraints such as infestation from pests that are threatening areca nut yield and the life of the trees, decreasing use among the general public, and the effects of climate change. Hence, it is imperative to find approaches that can assist areca growers to improve their yield and income. One of the most feasible and effective approaches to increase income from areca crops is to valorize the byproducts and coproducts generated during areca processing and reduce or eliminate disposal by burning or burying. The byproducts and coproducts generated during areca processing also offer a sustainable and environmentally friendly resource to develop bioproducts for various applications. This book attempts to provide detailed information on the availability, properties, and prospects of utilizing areca byproducts and coproducts toward developing value-added bioproducts and biomaterials.

Table 1.4 Features of products made using areca nuts in different countries (Luong 2023)

Product example	Ingredients	Processing methods	Region of interest
Dried ripe nut (Challi) Kottapak, Kallipak	Whole ripened areca nut	Sun dry, kernel removal	India
Scented suparis (gutka)	Matured Areca nut	Dehusk, cut into pieces, boil, coat with concentrated extract from boiling areca nuts, dry	India
Pan masala	Areca nut, catechu, cardamom, lime, natural and artificial scented or flavoring additives (herbs or spices, flowers) with or without tobacco	Dry, coat, mix	India
Zarda	Areca nut, tobacco leaves, slaked lime, fragrance, spice, and flavors	Boil tobacco leaves in water with lime and spice until evaporated, mix with crushed areca,	India
Qinggao	Areca nut husk	Coat with syrup, add sesame seeds or Chinese wolfberry	Hunan, China
Yanguo	Areca nut husk	Smoke, coat with syrup, and add hydrated lime	Hunan, China
Processed areca nut	Areca nut	Boil, pour, cook, bake, steam, marinate, brine, dry, and so forth	Hunan, China
Areca nut	Unprocessed areca nut	Fresh (unripe) nut or mature nut	Taiwan, China
Lao-hwa	Areca nut, betel inflorescence, slaked lime	Fresh (unripe) nut and slaked lime wrapped in betel inflorescence	Taiwan, China
Betel quid	Areca nut, betel leaf, slaked lime	Fresh (unripe) nut and slaked lime wrapped in betel leaf	Taiwan, China
Processed areca nut	Areca nut and additives for flavor, such as slaked lime, orange peel, spices, catechu, etc.	Boil, cook, or dry and mix with added components	Taiwan, China

Fig. 1.9 Arecanut husks disposed of as waste

Fig. 1.10 Areca sheath attached to the tree (left) and sheaths collected and dried (right) for further processing

1.6 Conclusions

Cultivation and processing of areca have been practiced for centuries but could decline and face considerable challenges. As with other crops, areca cultivation generates substantial residues that are at present of limited use or value. However, the transition towards a circular and green economy and the compulsions to reduce the use of synthetic polymer-based products will make the use of crop residues inevitable. Utilizing the byproducts and coproducts generated during areca processing is a technically and economically feasible option to replace the synthetic polymer products in current use.

Chapter 2
Chemical, Physical Properties, and Applications of Areca Nuts

2.1 Growth and Composition of Areca Nuts

Areca (*Areca catechu* L.) is a fruit typically grown under tropical conditions and used for various food and non-food applications. Although the primary produce areca nuts (Fig. 2.1), is considered one of the most common psychoactive substances similar to tobacco, alcohol, and coffee, the nuts have several bioactive and pharmaceutical components and are hence considered useful for several medical applications (Fig. 2.2).

In terms of morphological structure, the fruit can be divided into six major parts as shown in Fig. 2.3. Further, the main chemical constituents in areca nuts can be classified into four groups, i.e., alkaloids, polyphenols, tannins, and other trace elements (Fig. 2.5) (Luong 2023; Dangore-Khasbage 2023; Yuan 2018; Sun 2024). Although considered to have carcinogenic compounds, areca nuts have diverse uses and are processed into various shapes and sizes and included in products with different features and product names (Luong 2023). Although the nuts contain more than 60 constituents (Table 2.1) the alkaloids arecoline, arecaidine, guvacoline, and guvacine and their derivatives (Figs. 2.4 and 2.5) (Jain 2013) have been reported to be carcinogenic. The extent or amounts of carcinogenic compounds vary between species, location, and other parameters and should be monitored carefully. In commercial products, the total alkaloid levels vary between 4.48 and 6.5 mg/g for products in India and slightly higher at 7.2 mg/g for products sold in China (Jain 2017). The components of areca nut were also found to vary considerably even within a small geographical region (Table 2.2). Recently, nearly 52 compounds were identified in areca nuts, which were not reported earlier for areca nuts grown in India or other countries (Hugar 2024). The alkaloid content in areca nuts was also reported to change with the age (maturity level) of the areca nuts. For instance, the guvacine content was 1871 μg/g in mature areca nuts compared to only 303 μg/g in the young green nuts, and total alkaloid content was 707 μg/g in mature nuts compared to 254 μg/g in the unmatured nuts (Gupta 2020). Similarly, the amount of alkaloids

N. Reddy and P. Aramwit, *Areca catechu and its byproducts*,
https://doi.org/10.1007/978-3-032-22134-6_2

Fig. 2.1 The plant, fruit, husk, and seed of *Areca catechu*. (A) Plant *A. catechu*. (B) Fresh and tender fruit. (C) Ripe fruit. (D) Arecae pericarpium. (E) Arecae semen. (F) Arecae semen tostum. (G) Areca husks(Sun 2024b). Published with permission through Creative Commons Attribution License (CC BY)

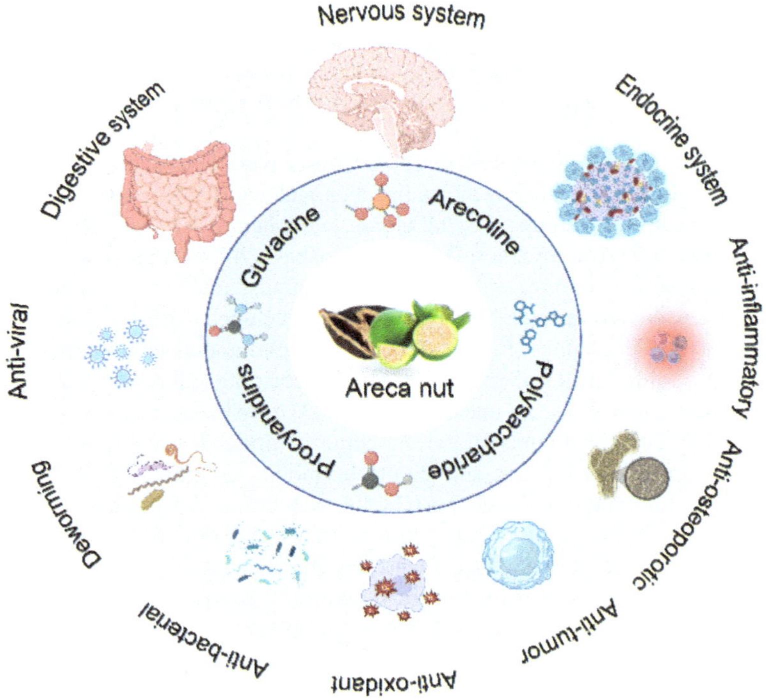

Fig. 2.2 Areca tree (A), areca fruits (B and C), nuts separated from the shell (E and F) and areca nut husks (D and G). (Sun 2024). Reproduced with permission through open access publishing

obtained from areca components varied considerably depending on the extraction conditions and parts of the areca plant used (Table 2.3) (Zhao 2025).

Similarly, substantial differences in the physiological parameters, texture, and flavor release behavior were noticed in areca nuts obtained from different regions. Samples collected from China (Hainan), Vietnam, Indonesia, and Myanmar showed substantial differences in their structure, physical, and chemical properties (Yang 2023). Fiber content and lignin were similar among the samples, whereas polyphenol (2.7–3.3 mg GAE/mL) content was slightly different (Yang 2023). Texture analysis showed substantial differences in the hardness, chewiness, and springiness among the different samples tested (Table 2.4). Such differences in texture were considered to contribute to the flavor release behavior, such as time taken, amounts of solids released, and extent of aroma from the nuts. As can be expected, the type of phytochemical compounds extracted from areca nuts and their activities also depend on the type of solvent used and the conditions during extraction. An attempt was made to use deep eutectic solvents and study the properties of the extracted components. The total phenolic content, total flavonoid content, and total alkaloid content extracted from areca nuts varied between 0.74 to 57.9 mg/g, 4.2 to 428 mg/g, and 0.6 to 33.9 mg/g, respectively. The phytochemical compound levels also varied considerably depending on the solvent used, as seen from (Table 2.5). The extracted components showed high bioactivities, including antioxidant, glucosidase inhibitory, and antimicrobial properties, preferable for pharmaceutical and other applications (Wang 2021).

2.2 Areca Nuts and Fibrosis

Aareca nuts are notoriously known for their fibrosis and cancer-causing effects. The International Agency for Research on Cancer. (IARC) has classified areca nuts among group 1 carcinogens. Several studies have linked betel quid chewing with oral cancer, particularly in South Asia and the Pacific regions (Warnakulasuriya 2022). Areca, in combination with betel quids and/or tobacco are commonly consumed in many countries and linked to cancer and other diseases. The alkaloids in areca nuts, particularly arecoline is considered to be the most potent cancer-causing chemical (Tables 2.6 and 2.7). Use of areca nuts leads to chronic inflammation, increased collagen synthesis, collagen crosslinking, and decreased collagen degradation, all of which contribute to Oral Submucous Fibrosis (OSF) and oral squamous cell carcinoma (Fig. 2.6) (Liu 2024) (Table 2.8). The major steps involved in the chronic inflammatory process are shown in Fig. 2.7 (Prabhu 2014). Arecoline, the major constituent of areca, has been reported to cause several harmful effects in humans and animals and adversely affect various organs (Huang 2024). Arecoline was found to be extensively hydrolyzed to arecaidine in human liver with an apparent intrinsic clearance of 57.8 ml/min/kg, considerably higher than 11.6 mL/min/kg for guvacoline, compared to 0.654 ml/min/kg and 0.466 ml/min/kg, respectively, in cytosol (HLC) (Paolillo 2025). Arecoline decreased the number of implanted embryos, and concentrations higher than 0.4 mM

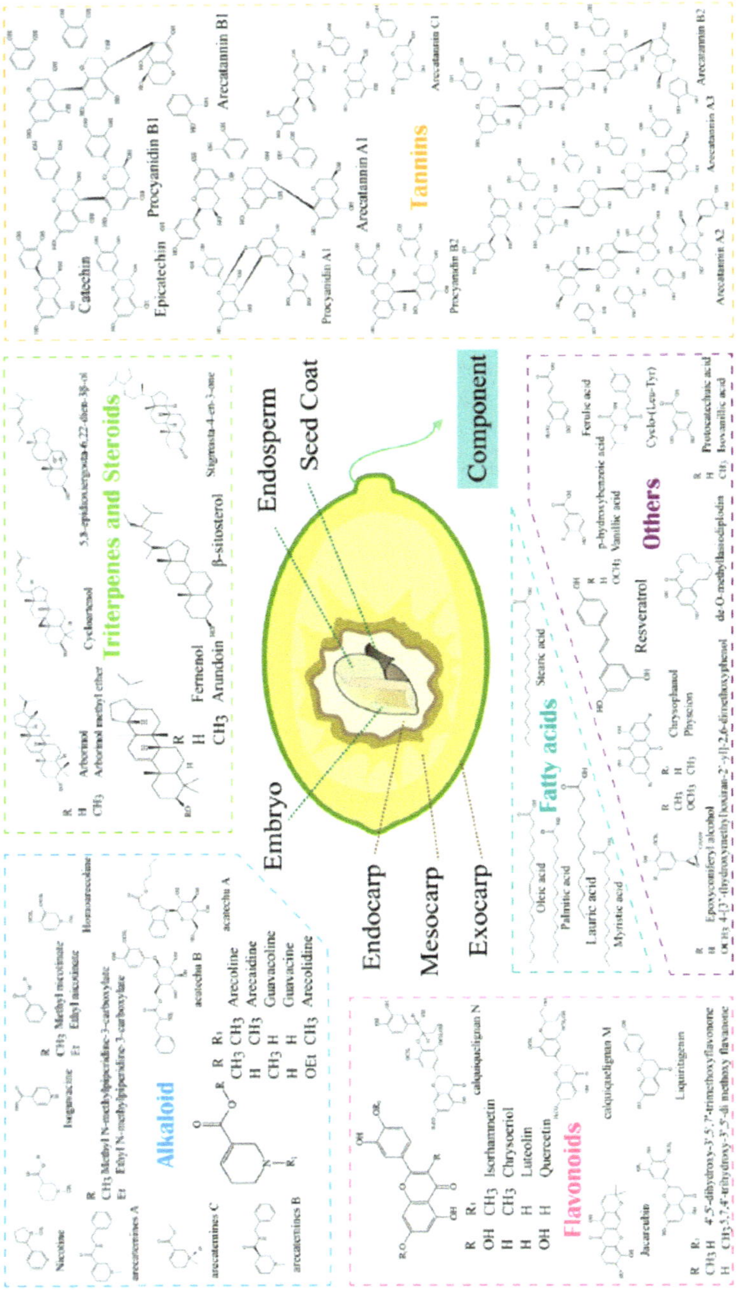

Fig. 2.3 Structure and chemical composition of areca nut. Areca nut consists of pericarp (exocarp, mesocarp, and endocarp) and seed (seed coat, endosperm, and embryo). The chemical components identified in areca nut mainly include alkaloids, flavonoids, tannins, fatty acids, triterpenes, and steroids (Sun 2024). Reproduced with permission through open access publishing

Table 2.1 The major compounds identified from the ethanolic extract obtained from unripe areca nut separated through UHPLC–MS/MS (Hugar 2024)

No	tR (min)	Compound name	Classification	Formula	Calculated mass	m/z	Exact mass	Mass accuracy mDa	Base peaks (%)
1	0.62	Brucein D	Quassinoid	$C_{2}0H_{26}O_{9}$	410.16	411.17	410.16	1.4	0.14
2	0.63	Syringic acid	Phenolic acid	$C_9H_{10}O_5$	198.05	199.06	198.05	−3.3	0.28
3	0.72	2-hydroxypalmitic acid	Fatty acid	$C_{16}H_{32}O_3$	272.24	273.24	272.24	1.0	0.07
4	0.78	Arecoline	Alkaloid	$C_8H_{13}NO_2$	155.09	156.10	155.09	−0.7	18.01
5	0.81	Lindenene	Terpene	$C_{15}H_{18}O$	214.14	215.14	214.14	−0.2	0.48
6	1.92	Procyanidin B2	Proanthocyanidins	$C_{30}H_{26}O_{12}$	578.14	577.13	578.14	−0.7	1.46
7	1.94	2,5-Dihydroxybenzoic acid	Phenolic acid	$C_7H_6O_4$	154.03	155.02	154.02	−7.6	0.08
8	2.27	Arecatannin B1	Proanthocyanidins	$C_{45}H_{38}O_{18}$	866.21	867.21	866.20	−1.5	28.25
9	2.34	Decursinol	Caumarin	$C_{14}H_{14}O_4$	246.09	245.08	246.09	−0.6	0.14
10	2.34	Nubigenol	Chalcone	$C_{15}H_{14}O_6$	290.08	289.07	290.08	−0.2	4.29
11	2.35	Onjixanthone II	Xanthone	$C_{15}H_{12}O_7$	304.06	303.05	304.06	−0.9	0.19
12	2.38	3-Hydroxybenzoic acid	Phenolic acid	$C_7H_6O_3$	138.03	139.04	138.03	−0.1	1.34
13	2.38	1-Hydroxy-3,7-dimethoxyxanthone	Xanthone	$C_{15}H_{12}O_5$	272.07	273.08	272.07	−0.8	0.41
14	2.38	(+)-Catechin	Flavan-3-ol	$C_{15}H_{14}O_6$	290.08	291.09	290.08	−0.1	19.09
15	2.40	Plathymenin	Flavanone	$C_{15}H_{12}O_6$	288.06	289.07	288.06	−0.8	0.28
16	2.91	6-Methoxyeriodictyol	Flavanone	$C_{16}H_{14}O_7$	318.07	319.08	318.07	0.3	0.07
17	3.62	1,8-Dihydroxy-2,3,4,5-tetramethoxyxanthone	Xanthone	$C_{17}H_{16}O_8$	348.08	349.09	348.08	−1.5	0.10
18	4.00	Propelargonidin	Proanthocyanidins	$C_{30}H_{26}O_{12}$	578.14	579.15	578.14	−0.3	4.35
19	4.26	5,7,2′,3′-Tetrahydroxyflavone	Flavone	$C_{15}H_{10}O_6$	286.05	287.05	286.05	−1.1	0.09
20	4.44	3,4-Dihydroaucubin	Iridoid glycoside	$C_{15}H_{24}O_9$	348.14	347.13	348.14	−0.2	0.65

(continued)

Table 2.1 (continued)

No	tR (min)	Compound name	Classification	Formula	Calculated mass	m/z	Exact mass	Mass accuracy mDa	Base peaks (%)
21	4.75	Isorhamnetin	Flavonol	$C_{16}H_{12}O_7$	316.06	317.07	316.06	−0.1	0.85
22	4.75	Isorhamnetin-3-O-glucoside	Flavonol glycosides	$C_{22}H_{22}O_{12}$	478.11	479.12	478.11	0.0	0.16
23	4.75	Complanatuside	Flavonol glycosides	$C_{28}H_{32}O_{16}$	624.17	625.18	624.17	0.8	2.20
24	4.95	Macranthol	Lignan	$C_{27}H_{26}O_3$	398.19	397.18	398.19	0.3	0.12
25	4.97	Phloretin	Chalcone	$C_{15}H_{14}O_5$	274.08	275.09	274.08	−0.3	0.18
26	4.97	(−)-Epiafzelechin	Flavan-3-ol	$C_{15}H_{14}O_5$	274.08	275.09	274.08	−0.3	0.18
27	5.30	Isoferulic acid	Phenolic acid	$C_{10}H_{10}O_4$	194.06	193.05	194.06	−0.6	0.25
28	5.60	Methyl rosmarinate	Phenolic acid	$C_{19}H_{18}O_8$	374.10	375.11	374.10	−0.1	0.07
29	5.82	Thymonin	Flavone	$C_{18}H_{16}O_8$	360.08	361.09	360.08	0.3	0.33
30	5.85	Ombuin	Flavonol	$C_{17}H_{14}O_7$	330.07	331.08	330.07	0.2	0.18
31	6.01	Hemiariensin	Phenolic acid	$C_{22}H_{24}O_7$	400.15	401.16	400.15	0.3	0.24
32	6.30	Glabrone	Isoflavonoids	$C_{20}H_{16}O_5$	336.10	335.09	336.10	1.1	0.11
33	6.44	Kanzonol Z	Flavanones	$C_{25}H_{26}O_5$	406.18	405.17	406.18	0.4	0.09
34	6.44	Kuwanon U	Flavanones	$C_{26}H_{30}O_6$	438.20	437.20	438.21	1.0	1.43
35	6.73	Artonin E	Flavone	$C_{25}H_{24}O_7$	436.15	435.15	436.15	1.2	0.10
36	6.77	1-Hydroxy-2,3,4,5-tetramethoxyxanthone	Xanthone	$C_{17}H_{16}O_8$	332.09	333.10	332.09	−0.6	0.07
37	6.94	Coniferyl alcohol	Monolignol	$C_{10}H_{12}O_3$	180.08	179.07	180.08	−0.3	4.45
38	6.97	Tenaxin I	Flavone	$C_{18}H_{16}O_7$	344.09	345.10	344.09	−0.7	0.56
39	6.98	Ethyl caffeate	Phenolic acid	$C_{11}H_{12}O_4$	208.07	209.08	208.07	−1.0	0.09
40	7.66	Kadsulignan F	Lignan	$C_{29}H_{32}O_{11}$	556.19	557.20	556.19	−0.4	0.10
41	7.85	Glycyphyllin	Chalcones	$C_{21}H_{24}O_9$	420.14	421.15	420.14	−0.8	0.07

(continued)

Table 2.1 (continued)

No	tR (min)	Compound name	Classification	Formula	Calculated mass	m/z	Exact mass	Mass accuracy mDa	Base peaks (%)
42	7.85	Rhapontin	Stilbin	$C_{21}H_{24}O_9$	420.14	421.15	420.14	−0.8	0.07
43	8.02	2′,4′-Dihydroxy-7-methoxy-8-prenylflavan	Flavonoid	$C_{21}H_{24}O_4$	340.17	339.16	340.17	−0.1	0.10
44	8.76	5,3′-Dihydroxy-6,7,4′-trimethoxyflavanone	Flavanone	$C_{18}H_{18}O_7$	346.11	347.11	346.10	−1.2	6.36
45	10.75	Hydnocarpin	Flavone	$C_{25}H_{20}O_9$	464.11	463.11	464.11	1.6	0.24
46	10.82	Glycyrrhizol B	Isoflavone	$C_{21}H_{18}O_5$	350.12	349.11	350.12	0.2	0.14
47	11.89	Kushenol D	Chalcone	$C_{27}H_{32}O_6$	452.22	451.21	452.22	0.9	0.14
48	12.01	Schisantherin J	Lignan	$C_{35}H_{40}O_{11}$	636.26	635.25	636.26	0.2	0.38
49	12.19	Sanggenone D	Chalcones	$C_{40}H_{36}O_{12}$	708.22	707.21	708.22	−2.5	0.13
50	12.31	Kurarinol	Flavanone	$C_{26}H_{32}O_7$	456.21	455.21	456.22	0.9	0.09
51	12.32	Archangelicin	Caumarins	$C_{24}H_{26}O_7$	426.17	425.16	426.17	2.0	0.47
52	12.41	Mulberrofuran O	Chalcones	$C_{39}H_{34}O_9$	646.22	645.21	646.22	1.9	0.38
53	13.05	Kushenol L	Flavonols	$C_{25}H_{28}O$	440.18	439.18	440.18	−0.2	0.33
54	13.14	Glabrolide	Terpenoid	$C_{30}H_{44}O_4$	468.32	467.32	468.32	−1.2	0.10
55	13.23	Garcinone D	Xanthone	$C_{24}H_{28}O_7$	428.18	427.18	428.19	1.9	0.30
56	13.41	Sanggenol A	Flavanone	$C_{25}H_{28}O_6$	424.19	423.18	424.19	−0.1	0.07
57	14.33	Canthaxanthin	Carotenone	$C_{40}H_{52}O_2$	564.40	563.39	564.40	2.1	0.13
58	14.55	Kadsutherin	Lignan	$C_{27}H_{32}O_7$	468.21	467.21	468.21	−1.4	0.12
59	14.88	Kansenonol	Terpenoid	$C_{30}H_{48}O_3$	456.36	455.35	456.36	−0.9	0.10
60	14.94	Gancaonin F	Flavanone	$C_{21}H_{16}O_6$	364.09	365.10	364.10	1.8	0.07
61	15.11	Artemetin	Flavone	$C_{20}H_{20}O_8$	388.12	387.11	388.11	−1.6	0.16

Reproduced with permission through open access publishing

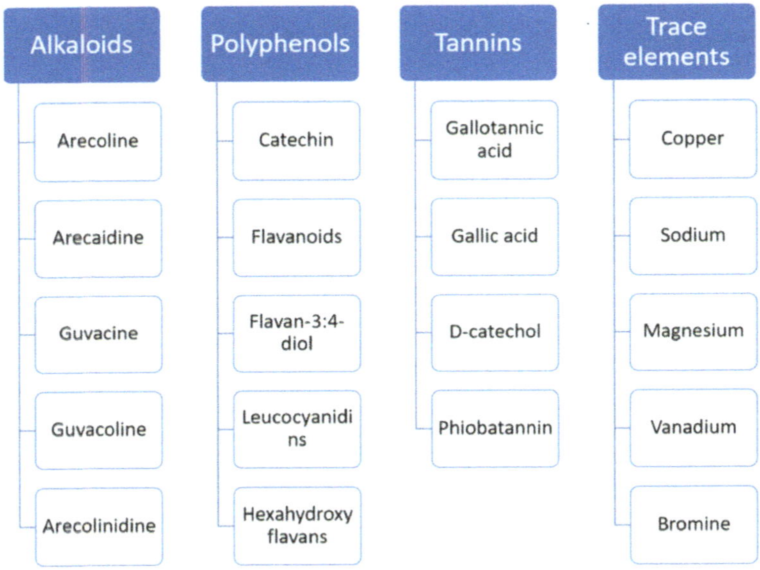

Fig. 2.4 Classification of the major chemical components in areca nuts (Dangore-Khasbage 2023). Reproduced with permission through open access publishing

were found to increase apoptosis and decrease viability of C_2C_{12} cells (Chang 2012). Such behavior was attributed to the inhibition of myogenic differentiation through the reduction of STAT3 activation during myotube formation (Jeng 2001). Further studies have shown that arecoline causes clustering of acetylcholine receptors in the postsynaptic membrane at the neuromuscular junction (NMJ and by destabilizing the agrin-induced or spontaneous AChR cluster formation and decreasing the expression of myogenin in C_2C_{12} myotubes (Fig. 2.8). Such effects are suggested to lead to muscle fatigue due to habitual areca nut chewing (Chang 2013). Betel nut alkaloids, particularly arecoline, have also been found to cause obesity and metabolic syndrome. Tian et al. reported that arecoline and not arecaidine or guvacine were able to arrest preadipocyte growth, decrease cyclin-dependent kinase 1, increase p53 and cyclin B1 protein, and inhibit AMP-activated protein kinase activity through an intracellular ROS pathway, thereby affecting body weight, fat cells, and metabolic syndrome (Tian 2018).

A study was done to establish the relationship between areca nut chewing and metabolic syndrome (Shafique 2013). The study revealed a strong relationship between areca nut chewing and metabolic syndrome, with the effect being more pronounced in individuals using tobacco additives along with the areca nuts (Table 2.9). Studies have further revealed that systemic inflammation is the most probable pathway that would increase the risk of systemic diseases among areca nut chewers (Shafique 2012). In another study, it has been suggested that the use of areca nuts increased the risk of cardiovascular diseases and cancers due to the

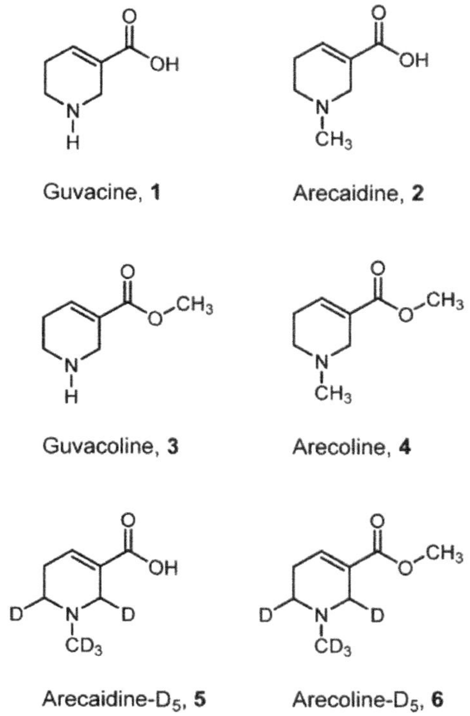

Fig. 2.5 Structure of the different alkaloids found in areca nuts (Jain 2017). Reproduced with permission from American Chemical Society

increase in levels of circulating inflammatory markers that would lead to chronic inflammation and oral cancer and multiple diseases (Shafique 2012). A proposed mechanism for the cause of inflammation through the generation of reactive oxygen species is shown in Fig. 2.9 (Senevirathna 2023). It was concluded that complex mechanisms are involved in the process of oral carcinogenesis attributable to the chewing of areca nuts. Other researchers have suggested multiple cytotoxic effects, including inflammation, tissue hypoxia, DNA damage, autophagy, etc., that lead to carcinogenesis. Generation of ROS further leads to activation of signaling pathways, epithelial-mesenchymal transition, and eventually cancer stemness conversion (Fig. 2.10) (Li 2019; Lu 2010).

In a detailed study to understand the effects of areca nut chewing on oral submucous fibrosis and oral squamous cell carcinoma, it was found that areca nut extracts impair T-cell activation and cause production of tumor necrosis factor-α (TNF-α) and interleukin (IL)-6 in oral epithelial cells. Further, areca nut extracts downregulate the production of alkaline phosphatase and upregulates the production of Receptor Activator of Nuclear Factor Kappa-B Ligand (RANKL) in osteoblasts. Specifically, areca nut extracts neutralize and inhibit the bactericidal properties of neutrophilis in the oral cavity, subsequently leading to apoptosis (Ho 2010; Chang 2013), the extent of which depends on the concentration of the areca nut extracts (Fig. 2.11). Exposure of areca nut extracts to HSC-2 and HSC-3 cells leads to early and late apoptosis and

Table 2.2 Variations in the TPC, TFC, and arecoline contents extracted from areca nuts grown in Karnataka, India (Hugar 2024)

Sample no	TPC (mg GAE/g FW)	TFC (mg CE/g FW)	% of TFC in TPC	Arecoline content (mg/g FW)
0	66.02 ± 0.003	52.98 ± 0.003	80.25	1.36 ± 0.09
1	87.16 ± 0.011	66.35 ± 0.007	76.12	1.62 ± 0.23
2	84.55 ± 0.018	59.65 ± 0.018	70.55	1.76 ± 0.14
3	48.52 ± 0.015	44.52 ± 0.005	91.76	0.86 ± 0.02
4	59.83 ± 0.002	52.86 ± 0.006	88.35	2.17 ± 0.13
5	61.25 ± 0.005	53.18 ± 0.007	86.82	0.71 ± 0.07
6	92.73 ± 0.011	89.60 ± 0.004	96.62	3.48 ± 0.12
7	96.82 ± 0.006	78.06 ± 0.004	80.62	1.33 ± 0.13
8	107.61 ± 0.014	91.43 ± 0.008	84.96	2.14 ± 0.03
9	105.68 ± 0.003	84.23 ± 0.007	79.70	2.39 ± 0.20
10	109.94 ± 0.005	101.37 ± 0.010	92.20	2.89 ± 0.13
11	76.25 ± 0.005	73.81 ± 0.001	96.80	2.22 ± 0.27
12	165.29 ± 0.005	158.13 ± 0.013	95.66	3.29 ± 0.20
13	99.609 ± 0.002	88.75 ± 0.018	89.11	1.98 ± 0.07
14	129.91 ± 0.002	56.0 ± 0.007	43.11	3.66 ± 0.03
15	120.13 ± 0.011	91.5 ± 0.013	76.16	2.02 ± 0.27
16	104.64 ± 0.005	86.81 ± 0.003	82.96	2.5 ± 0.02
17	145.45 ± 0.003	54.09 ± 0.001	37.18	2.55 ± 0.14
18	129.48 ± 0.005	109.32 ± 0.003	84.43	2.98 ± 0.03
19	88.21 ± 0.002	81.92 ± 0.002	92.86	1.91 ± 0.27
20	113.56 ± 0.011	78.86 ± 0.007	69.44	2.89 ± 0.20

Reproduced with permission through open access publication

was hypothesized to be due to the caspace-3 enzyme activity, both through intrinsic and extrinsic pathways (Sari 2019).

The carcinogenic effects of areca nut consumption alone or in combination with tobacco or betel quids are well documented. In addition to these, areca nut consumption is also linked to reproductive issues in humans and animals (Kumar 2013). Consumption of areca nut with or without smokeless tobacco was found to adversely impact pregnant women (Table 2.10) (Kumar 2013). Studies have shown that continuous use of tobacco develops dependence and causes significant health implications (Benegal 2008). Recent studies have shown that areca nut extract modulates the immune system during oral carcinogenesis by upregulating mitochondrial metabolism and p-CREB/VCAM-1 signaling in macrophages, decreased mitochondrial metabolism and increased EMT and migration in oral precancer cells. A highly positive correlation was established between M2 macrophages and oral cancer progression with areca nut extract-treated macrophages promoting migration and mesenchymal phenotypes in oral precancer cells. It was specifically shown

Table 2.3 Different extraction methods are used to obtain alkaloids from parts of the areca plant (Zhao 2025)

Extraction method	Extract object	Optimal parameters	Extraction rate (mg/g)
Solvent extraction	Arecoline in AN waste	The particle size is 60 mesh, the liquid-to-solid ratio is 15:1, the extraction time is 45 min, the extraction time is 2, the extraction temperature is 80 °C, and the mass concentration of hydrochloric acid is 0.22%	8.330 ± 0.109
Heat reflux extraction	Arecoline in AN inflorescence	The liquid-to-solid ratio is 80:1, the extraction time is 25 min, and the extraction time is 3	4.31 ± 0.047
Ultrasonic-assisted extraction	Arecoline in AN husk	The concentration of ethanol is 85%, the liquid-to-solid ratio is 8:1, the extraction temperature is 65 °C, the pH is 8	1.987
Supercritical CO_2 extraction	Arecoline in AN husk	The extraction pressure is 33 MPa, the extraction temperature is 63 °C, and the microwave power is 480 W	4.48 + 0.07
SWE extraction	Areca alkaloids in AN seed	The extraction time is 43 min, the extraction temperature is 143 °C, and the liquid-to-solid ratio is 35:1	4.23
SWE extraction	Areca alkaloids in AN husk	The extraction temperature is 110 °C, the liquid-to-solid ratio is 19:1, the extraction time is 50 min	8.17

Table 2.4 Variations in the textural parameters for the areca nuts obtained from different locations (Yang 2023)

	Hardness/g	Chewiness/g	Springiness
AN1	9396.36 ± 2184.87	1422.44 ± 413.30	0.3006 ± 0.0364
AN2	10,452.56 ± 4251.33	1598.52 ± 455.38	0.3119 ± 0.0382
AN3	12,554.55 ± 3952.30	1648.65 ± 513.32	0.3165 ± 0.0389
AN4	11,855.95 ± 3125.41	1552.34 ± 486.27	0.3163 ± 0.0375
CAN1	6069.00 ± 2537.69	1025.61 ± 305.80	0.2157 ± 0.0242
CAN2	6064.66 ± 1626.23	1035.29 ± 351.64	0.2189 ± 0.0218

Table 2.5 Contents of the arecanut seed extracts obtained using various solvents (Wang 2021)

Contents	H_2O	50% MeOH	50% EtOH	50% Acetone	EtAc	Pro-Gly	Pro-Ethy	Pro-LA
Phenolics and flavonoids, (μg/g DW)								
	80.37 ± 0.29	36.44 ± 0.20	77.29 ± 0.04	15.83 ± 0.07	5.26 ± 0.02	72.57 ± 0.05	111.58 ± 0.04	168.68 ± 0.04
Gallocatechin	488 ± 0.49	735 ± 0.17	298 ± 0.37	0	0	1087 ± 0.32	1042 ± 0.37	0
Epigallocatechin	16507 ± 0.19	20349 ± 0.40	11267 ± 0.32	20266 ± 0.34	0	14858 ± 0.37	22859 ± 0.23	23626 ± 0.38
D-catechin	438 ± 0.19	4462 ± 0.19	4346 ± 0.27	4822 ± 0.43	0	939 ± 0.40	2732 ± 0.39	0
Caffeic acid	2.95 ± 0.48	5.28 ± 0.26	104 ± 0.15	27 ± 0.19	0	0	0	274 ± 0.17
Epigallocatechin gallate	0	59.64 ± 0.17	0	0	0	0	0	0
Gallocatechin gallate	0	137 ± 0.16	0	0	0	0	0	0
Vanillin	0	6.5 ± 0.30	17.4 ± 0.35	17.4 ± 0.21	0	0	0	0
P-Coumarin	4.59 ± 0.19	9.66 ± 0.16	6.5 ± 0.21	10.4 ± 0.27	0	0	0	0
Ferulic acid	0	10.21 ± 0.21	9.27 ± 0.27	0	0	0	0	0
Quercetin-3-O-galactoside	23.38 ± 0.22	21.49 ± 0.15	83 ± 0.18	7.4 ± 0.18	0	0	17.7 ± 0.11	0
Kaempferol-3-O-galactoside	20 ± 0.24	60.32 ± 0.19	104 ± 0.229	77.3 ± 0.11	31 ± 0.22	53.7 ± 0.19	55.5 ± 0.08	69 ± 0.17
Isorhamnetin-3-O-galactoside	1884 ± 0.70	6731 ± 0.31	10,104 ± 0.27	7893 ± 0.09	0	5117 ± 0.16	10,401 ± 0.24	5831 ± 0.27
o-Coumaric acid	0	0	0	0	0	0	34.6 ± 0.29	0
Quercetin	0	20 ± 0.13	22.7 ± 0.15	0	26.6 ± 0.26	0	0	0
Naringenin	0	5.3 ± 0.12	3.94 ± 0.15	4.3 ± 0.16	73.7 ± 0.21	0	0	109 ± 0.08
Kaempferol	0	0	7.22 ± 0.19	9.44 ± 0.26	0	0	0	0

(continued)

Table 2.5 (continued)

Contents	H_2O	50% MeOH	50% EtOH	50% Acetone	EtAc	Pro-Gly	Pro-Ethy	Pro-LA
Phenolics and flavonoids, (μg/g DW)								
Alkaloids (mg/g DW)								
Arecoline	3.3 ± 0.05	9.8 ± 0.01	8.55 ± 0.01	5.83 ± 0.01	0.43 ± 0.01	19.6 ± 0.08	35.86 ± 0.33	34.07 ± 0.62
Arecanine	1.48 ± 0.05	1.51 ± 0.04	1.90 ± 0.06	0.84 ± 0.03	0.20 ± 0.04	1.13 ± 0.07	2.11 ± 0.05	2.25 ± 0.06
Guvacoline	1.10 ± 0.04	4.82 ± 0.11	3.57 ± 0.10	2.53 ± 0.09	0.34 ± 0.01	7.53 ± 0.27	11.12 ± 0.35	15.89 ± 0.35
Guvacine	0.95 ± 0.04	3.06 ± 0.06	2.43 ± 0.04	1.52 ± 0.10	0.18 ± 0.02	0.53 ± 0.04	0.85 ± 0.04	0.78 ± 0.04

Reproduced with permission from Elsevier

Table 2.6 Effect of arecoline on different organs (Huang 2024)

Num	Action object	Experimental model	Experimental result	Arecoline dose
1	Oral system	Human umbilical vein endothelial cell	Induced endothelial–mesenchymal transformation	5–20 μg/mL
2	Oral system	EAhy926 endothelial cells	Promoted oral submucosal fibrosis	>0.4 mM
3	Cerebral nerve	Sprague–Dawley rat/mouse	Induced neuronal cell death	50–200 μM
4	Cerebral nerve	Mouse	Helped regulate lipid metabolism and reduce depressive behaviors	6 mg/kg
5	Cerebral nerve	Adolescent and adult male C57BL/6 J mice	Addiction	Adolescent mice: 0.03 mg/kg dose; Adult mice: 0.1 mg/kg dose
6	Cerebral nerve	Adult zebrafish	Anti-anxiety behavior	10 mg/L
7	Cerebral nerve	Zebrafish larvae	Induced hyperactivity	0.1 ppm
8	Cerebral nerve	Mouse	Enhanced learning and memory	500 mg/kg of wet Areca catechu extract
9	Heart and circulatory system	SD mouse	It can protect against myocardial damage caused by acute alcoholism	5 mg/(kg d)
10	Heart and circulatory system	Sprague–Dawley mouse	It has been demonstrated that the induction of apoptosis in the heart can be achieved through the stimulation of death receptors and mitochondrial-dependent apoptosis pathways	Low dose: 5 mg/kg/day and high dose: 50 mg/kg/day
11	Heart and circulatory system	Male BALB/c mice	Promoted defecation function, enhanced gastrointestinal peristalsis	30 mg/kg
12	Heart and circulatory system	C57BL/6 mice	Caused intestinal inflammation, intestinal damage, and changes in gut microbes	6 and 30 mg/kg

(continued)

Table 2.6 (continued)

Num	Action object	Experimental model	Experimental result	Arecoline dose
13	Embryo	Zebrafish embryos	Poor embryo survival, stunted growth	0.001–0.04%
14	Embryo	Pregnant mouse	In early pregnant mice, a reduction in the number of implanted embryos was observed, which resulted in the inhibition of trophoblast growth and expansion of the blastula	6.11 mg/kg/d
15	Hepatic system	Rat hepatocytes	Liver fibrosis and hepatocellular carcinoma	0.5 mM

Reproduced with permission through open access publishing

that arecoline induces M2 macrophage differentiation and secretion of oncogenic cytokines in the oral tumor microenvironment, promoting malignant transformation of oral premalignant cells. (Fig. 2.12) (Yuan 2024). In a study by Sharma et al., areca nut consumption was related to oral submucous fibrosis (OSF) due to the upregulation of inflammatory cytokines such as TGF-β which triggers the conversion of fibroblasts to myofibroblasts, characterized by α-SMA and γSMA expression. Arecoline was found to enhance the availability of reactive oxygen species (ROS), upregulate the expression of cytokines such as PAI-1, NF-κB, IGF-1, IL-13, and activate specific signaling pathways such as MEK, COX2, PI3K, which would lead to fibrosis. Further, adding lime during the chewing of areca would enhance the production of arecoline and hence lead to the severity of fibrosis (Fig. 2.13) (Sharma 2024).

In addition to arecoline, researchers have suggested the presence of other spasmogenic constituents in areca nut, which cause gastrointestinal stimulation, excessive salivation along with central effects. A study by Gilani et al. suggested that the presence of cholinomimetic and acetylcholinesterase (AChE) in areca nut crude extracts caused dose-dependent spasmogenic effect in rabbits (Fig. 2.14) (Gilani 2004). Although arecoline is considered to be the major cancer-causing substance, several other areca nut extracts have also been reported to cause cancer in human and animal cells (Salehi 2020). The possibility of arecoline contributing to inflammatory disorders such as atherosclerosis was demonstrated using endothelial cells (EAHY). It was found that long-term exposure to arecoline could damage the vascular endothelium and lead to cardiovascular diseases and oral mucosal alterations. EAHY cells exposed to areconline showed higher attachment to mononuclear cells in a dose-dependent manner. Although arecoline inhibited the migration of EAHY cells (Fig. 2.15), they enhanced the adhesion of U937 cells. Based on these observations, it was suggested that cytotoxicity, migration, and mononuclear cell adhesion to endothelial cells could promote pathogenesis and cause impairment of vascular functions (Tseng 2014).

Table 2.7 Effect of arecoline on the central nervous system (Liu 2024)

Effect	Animal/cell	Specific effect	Pathway/mediators	Dose
Beneficial effects	Xenopus laevis oocytes	Anti-inflammatory activity	As a silent agonist of α7 nAChR, targeting and regulating intracellular signaling against inflammation and pain	/
	Glioblastoma cell lines (U373 and U87MG)	Interfere with the aggressiveness of malignant gliomas	Inhibition of intermediate conductance Ca^{2+}-activated K^+ channels	10 and 30 μM
	Zebrafish	Cluster disruption and increased social interaction	Increased norepinephrine, serotonin, and DOPAC levels decreased 5-hydroxyindoleacetic acid/serotonin level, and homovanillic acid/dopamine ratios	10 mg/L
	Zebrafish	Motor hyperactivity	Binds with multiple mAChRs (M_1–M_4) to induce hyperactivity	0.001, 0.01, 0.1, and 1 ppm
	Male Swiss albino mice	Antinociception	By activation of central muscarinic receptors	0.3–1 mg/kg ip
	Rat	Attenuated a time perception impairment induced by daily scheduled feeding	By modulating central cholinergic	10 mg/kg/d
	Rat	Anti-phenobarbital sodium-induced sleep time	Not mentioned	0.5 mg
	Male ICR mice	Shortened the duration of ethanol-induced sleep	Acts as a muscarinic agonist to relieve ethanol-induced central depression and intoxication	0.125–1.0 mg/kg, s.c
	CPZ mice	Attenuating memory impairment and demyelination	Acts as a muscarinic receptor 1 cholinergic agonist to improve cognition and promote myelination processes in the frontal cortex	2.5 or 5 mg/kg/d

(continued)

Table 2.7 (continued)

Effect	Animal/cell	Specific effect	Pathway/mediators	Dose
	Female BALB/c mice	Increased the activity of preactivated NK cells	By stimulating the secretion of corticotropin-releasing hormone and adrenocorticotropic hormone	1.5 mg/kg
	Male albino rats	Improved retrieval and memory storage in the stair maze	Not mentioned	0.5 mg/kg
	Human (Alzheimer)	Low-dose arecoline improved cognitive performance, and the highest dose impaired psychomotor activation	By modulating central cholinergic	1, 2, or 4 mg/h infusions 2 h
	Human (Alzheimer)	Improved memory	As a cholinergic agonist, maintaining patients' cholinergic steady-state	0.042–1.7 mg/h Infusion for 11–16 d
	Human (Alzheimer)	Improved cognition	As a muscarinic receptor agonist, regulating patients' cholinergic system	0.5, 1, 2, 4, 8, 16, 22, 28, 34, and 40 mg/d
Neurotoxicity	Primary cortical neuron	Induction of neuronal cell death	By attenuating antioxidant defense and enhancing oxidative stress	50–200 μM
	PC12 Cells	Apoptosis	By inducing endoplasmic reticulum stress, attenuating H_2S levels, CBS, and 3-MST protein expression	0.5–2 mM
	Drosophila melanogaster	Neurotoxic agent and affected the life cycle parameters	By reducing acetylcholinesterase and MAO, increasing caspase-3, caspase-9 activity, and oxidative stress	20, 40 and 80 μM
	Zebrafish	Dyskinesia	By increasing ROS, endoplasmic reticulum stress, apoptotic p53 signaling pathway	10 μM
	Male albino rats	Decreased correct responses and accelerated spontaneous decay of memory	Not mentioned	3.5 and 8 mg/kg

(continued)

Table 2.7 (continued)

Effect	Animal/cell	Specific effect	Pathway/mediators	Dose
Addiction-related	Xenopus laevis oocytes	Habitual use	By activating addiction-related nAChR activity, receptors containing α4, β2, α6, and β3 subunits	/
	Xenopus laevis oocytes	Addiction	By activating α4 nAChR	100 μM
	Zebrafish	Withdrawal syndrome-like responses	Not mentioned	1 mg/L
	Pregnant women	Exceptional adverse birth outcome	Not mentioned	Not mentioned

Reproduced with permission through open access publishing

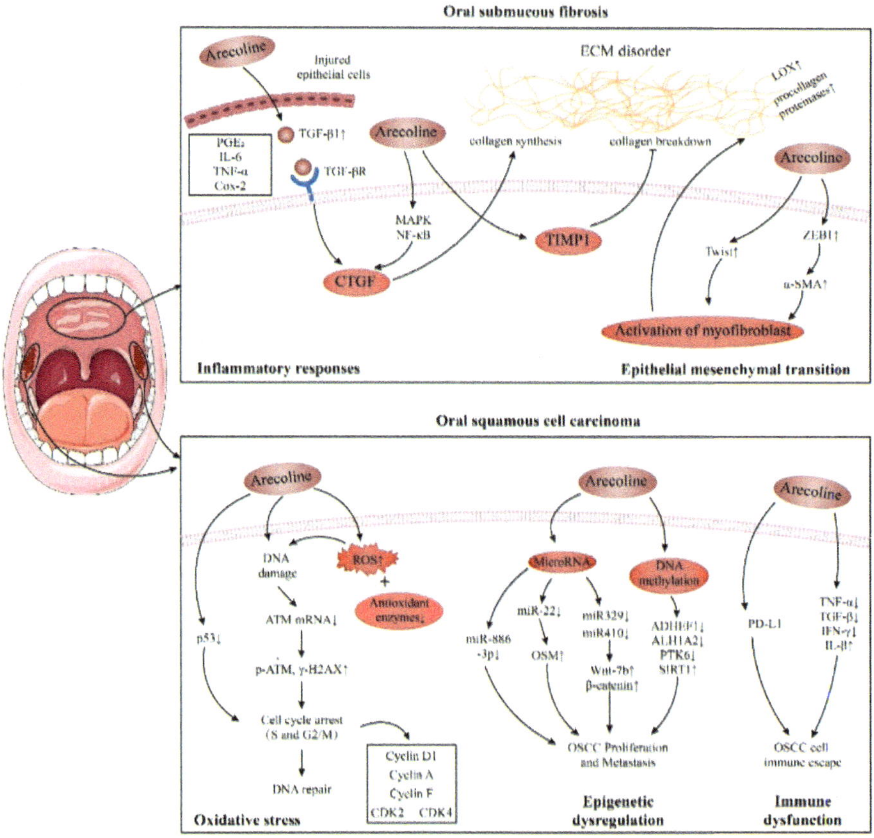

Fig. 2.6 Possible mechanisms of oral submucous fibrosis (OSF) and oral squamous cell carcinoma (OSCC) induced by arecoline (Liu 2024). Reproduced with permission through open access publishing

2.3 Beneficial Properties of Areca Nut Components

The bioactive compounds in areca nuts, such as various alkaloids, polyphenols, steroids, tannins, and fatty acids, are reported to provide antioxidant, antidiabetic, antiageing, and antiallergic properties and also have the ability to promote digestion (Fig. 2.16) (Guo 2024). Researchers have also suggested that the components in areca nuts have anti-viral properties and could be used for treating conditions in the nervous system, endocrine system, and tumors (Sun 2024). A summary of the bioactive compounds present in areca nuts and their beneficial aspects and mechanism of their biological activity are given in Table 2.11 (Sun 2024). However, the type, amount, concentration, and even activity of the compounds in areca nut depend on the species, climatic conditions, extraction and characterization methods used.

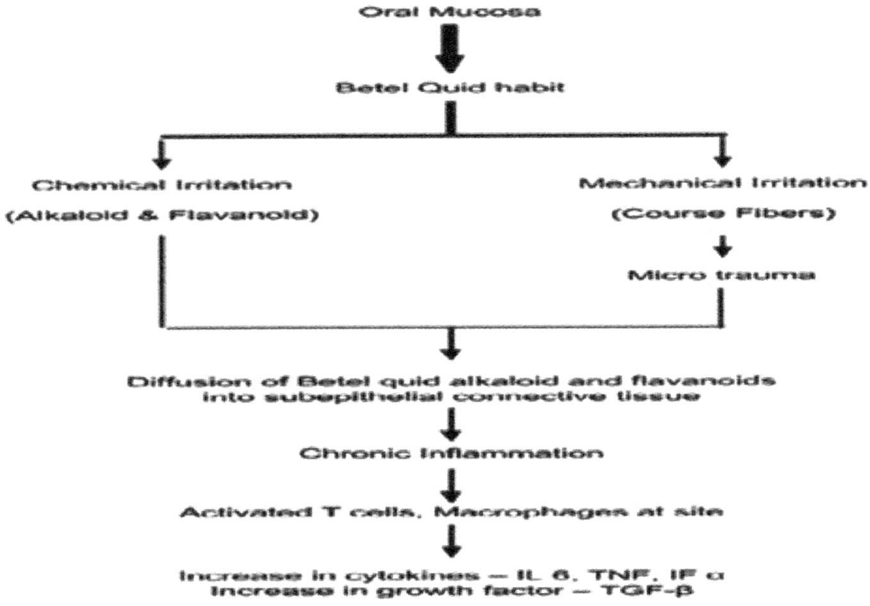

Fig. 2.7 Schematic representation of the chronic inflammatory process at the site of placement of the betel quid (Prabhu 2014). Reproduced with permission through open access publishing

It was reported that pH, extraction time, and substrate concentration affected the amounts of polyphenols extracted from areca nuts (Chavan 2013) (Tables 2.12 and 2.13). High amounts of polyphenols and flavonoids, such as catechin and epicatechin, were extracted from areca nuts using HPLC. Maximum level of arecoline was obtained when extraction was done using 80% acetone at pH 4 for 90 min and 10% arecanut concentration. Presence of catechin, epicatechin, and arecoline was confirmed through LC-MS, and the extracted phenols were considered to be useful for medical applications (Chavan 2013).

In addition to the regular components, several unique compounds such as triterpenoids and steroidals have been extracted from areca nuts (Tong 2024) (Table 2.14). Some of these compounds are suggested to provide beneficial effects, such as antidepressant activity by increasing serotonin and nonadrenaline levels in the brain. Similarly, areca nut extracts provide anti-inflammatory properties through the inhibition of cyclooxygenase (Tong 2024). Some of the beneficial pharmacological effects of areca nuts and their extracts are shown in Table 2.15 (Tong 2024), and some of the systemic effects of areca nuts are shown in Table 2.16 (Garg 2014).

Contrary to the widely perceived chronic and ill effects of areca nut consumption, several helpful medicinal properties of areca nuts have also been reported. For instance, areca nut components can be used as a veterinary medicine (anthelmintic agent), and the volatile components can be used as fragrances. Some of the tannins and fatty acids in areca nuts are reported to reduce cancer, inflammation, anxiety,

Table 2.8 Some of the toxicological and pharmacological effects of arecoline (Liu 2024)

Other toxicological and pharmacological effects of arecoline				
Effect	Animal/cell	Specific effect	Pathway/mediators	Dose
Respiratory system	Human and dermal and gingival fibroblast	Causing lung function impairment	In pro-inflammatory conditions (IL-4 and TNF-α), arecoline can induce eotaxin-1 release and alter the disease process in asthma	25 and 100 μg/mL
	Human	Asthma	Possibly related to arecoline-induced bronchoconstriction	/
Hepatotoxicity	Human liver microsome and Male Wistar rats	Hepatotoxicity	By increasing the hepatic CYP2E1 and CYP2B activity, induced oxidative damage, liver cirrhosis, and hepatocellular carcinoma	4, 20, and 100 mg/kg/d
	HA22T/VGH hepatoma cells	Inducing anoikis	By inhibiting STAT3 and SHP2 phosphorylation, decreasing the levels of anti-apoptotic factors, as well as by promoting the activity of pro-apoptotic factors	0–100 μg/mL
	Human and C57BL/6 mice's organ of Corti and spiral ganglions	Sensorineural hearing impairment	Reducing cochlear explant cell activity, inducing cell death, and ROS production by causing disruption of hair cells in the organ of Corti	0.2, 0.8, 2, and 10 mM
	Mice	Fatty degeneration and inflammatory infiltration	By increasing serum alkaline phosphatase, glutamate oxaloacetate transaminase, glutamate-pyruvate transaminase, and decreasing levels of reduced glutathione, glutathione-S-transferase, SOD, and catalase	10 mg/kg body weight

(continued)

Table 2.8 (continued)

Other toxicological and pharmacological effects of arecoline

Effect	Animal/cell	Specific effect	Pathway/mediators	Dose
	Mice	Decreasing nuclear size; the rough endoplasmic reticulum with profusely inflated cisternae and abundance of lipid droplets	By upregulating SGOT and SGPT (hepatotoxicity marker enzymes) in serum	5, 10, and 20 mg/kg body weight
Reproduction	Zebrafish embryos	Reducing survival of embryos with growth retardation and lower heart rate	General cytotoxic effects mainly due to intracellular thiol depletion	0.01–0.04% (wt/vol.)
	Oocyte	Apoptosis	By disrupting actin filament dynamics, spindle assembly, and kinetochore-microtubule attachment stability, mitochondrial distribution, and increasing oxidative stress levels	180 μg/mL
	ICR mice and blastocysts	Reduction of early embryos and inhibition of blastocyst growth and expansion	By inducing DNA damage, cell cycle arrest, or apoptosis	$0 - 8.47 \times 10^{-2}$ M
	Male rats	Stimulation of testosterone secretion	By activating L-type calcium channels, increasing 17β-hydroxysteroid dehydrogenase activity, and StAR expression, thereby stimulating testosterone production	1 μg/kg

(continued)

Table 2.8 (continued)

Other toxicological and pharmacological effects of arecoline				
Effect	Animal/cell	Specific effect	Pathway/mediators	Dose
Immunity and endocrine	Swiss albino mice	Lymphocyte depletion of the thymic cortex and the B and T lymphocyte areas in the spleen and MLN, Elevated corticosterone, SGOT, and SGPT levels, and decreased white and red blood cell counts	Not mentioned	20 mg/kg
	Adult male mice	The orientation of nuclei was irregular, follicle degeneration, a decrease in the T3, T4, number, and size of thyroid follicles, and an increase in the TSH level	MAChRs mediate the effect of arecoline on thyroid	10 mg/kg
	BALB/c mice	Reducing the spleen index, hemolysin, IL-2 production, and splenocyte proliferation induced by concanavalin A or lipopolysaccharide	Mediated via mAChRs	2 mg/kg
Fat	Mouse 3T3-L1 cells and human	Adipocyte dysfunction	Inhibiting adipogenic differentiation, inducing adenylate cyclase-dependent lipolysis, and interfering with insulin-induced glucose uptake	≥300 μM
	3T3-L1 cells	Regulating the growth of preadipocytes	Inhibiting the CDK family and the CKI pathway by inactivating AMPK activity as well as the intracellular ROS pathway	0–1,000 μ

Reproduced with permission through open access publishing

and platelet aggregation (Fig. 2.17) (Liu 2023). However, the pharmacological and toxicological effects of areca nut extracts are reported to differ in males and females (Fig. 2.18). As part of traditional medicine, areca nuts have been used to treat migraine due to the presence of procyanidins and polyphenolic compounds and their inhibitory activity on nitricoxidesynthase (iNOS). The antimigraine activity of the areca nut

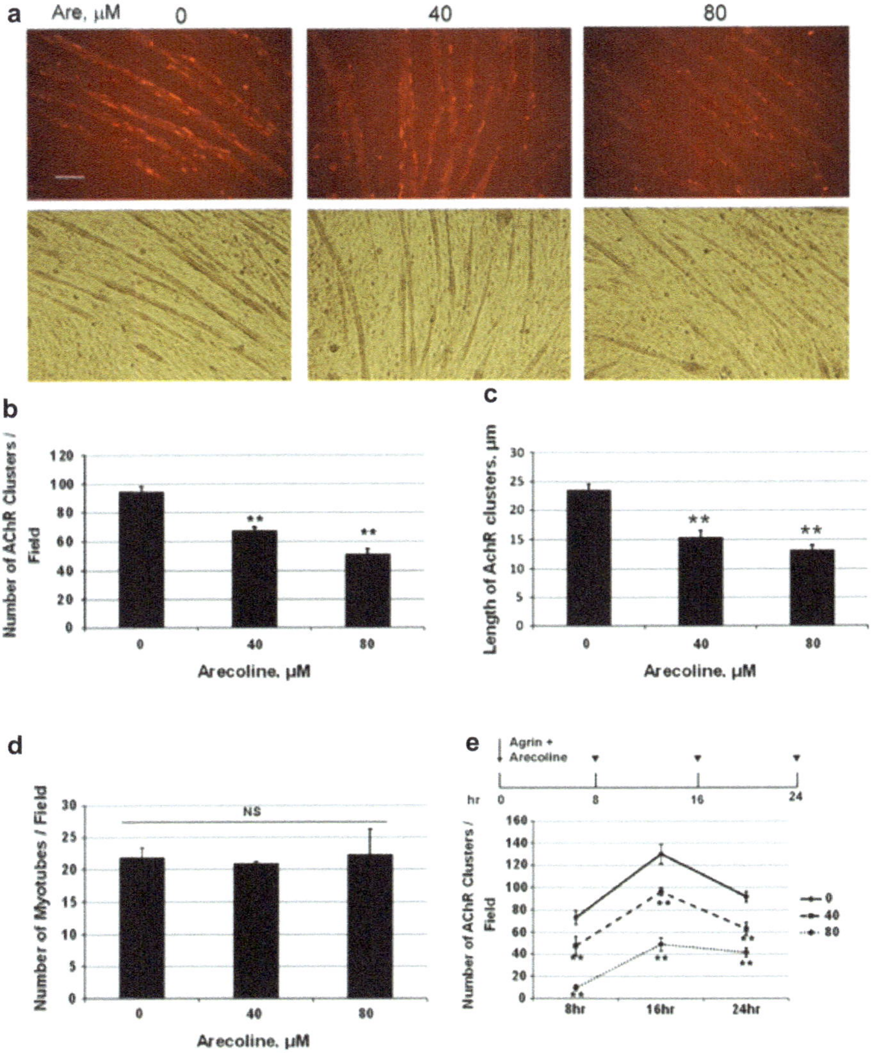

Fig. 2.8 The agrin-induced AChR clusters on C_2C_{12} cells were significantly decreased by 40 and 80 μM arecoline. **a** The C_2C_{12} cells were induced to form myotubes. Agrin stimulation, 0–80 μM arecoline treatment, and the acetylcholine receptor clustering assay were then performed. The AChR clusters were visualized by fluorescence microscopy (upper panels). Scale bar, 100 μm. The myotubes of the same field were also photographed by phase contrast microscopy (lower panels). **b** The number of AChR clusters from ten random fields under a fluorescence microscope was calculated after 24 h of arecoline and agrin treatment. **c** The length of 30 clusters from each treatment was measured. **d** The number of myotubes in the same fields was also counted. **e** The number of AChR clusters after 8, 16, and 24 h of arecoline and agrin treatment (Chang 2013). Reproduced with permission from Elsevier

Table 2.9 Relationship between areca nut chewing and metabolic syndrome (Shafique 2013)

Areca nut chewers				
	Non-chewers	Raw areca chewers	Areca with tobacco additives	P-value for trend
Total participants	722	231	117	
Central Obesity				
Obese, n(%)	231 (32.0)	94 (40.7)	57 (48.7)	
Odds ratio (95% CI)[a]	Reference	1.46 (1.07–1.98)	2.02 (1.36–3.00)	<0.001
Odds ratio (95% CI)[b]	Reference	1.42 (0.97–2.06)	1.65 (1.03–2.63)	0.015
Raised blood pressure				
Raised BP cases, n(%)	234 (32.4)	89 (38.5)	58 (49.6)	
Odds ratio (95% CI)[a]	Reference	1.31 (0.96–1.78)	2.05 (1.38–3.04)	<0.001
Odds ratio (95% CI)[b]	Reference	1.23 (0.89–1.70)	1.87 (1.25–2.81)	0.003
Reduced HDL				
Reduced HDL cases, n(%)	233 (32.8)	104 (45.0)	58 (50.0)	
Odds ratio (95% CI)[a]	Reference	1.72 (1.27–2.33)	2.06 (1.39–3.06)	<0.001
Odds Ratio (95% CI)[b]	Reference	1.55 (1.11–2.17)	1.66 (1.09–2.54)	0.003
Hypertrigylceridemia				
Hypertriglyceridemia cases, n(%)	150 (20.8)	72 (31.2)	41 (35.0)	
Odds ratio (95% CI)[a]	Reference	1.73 (1.24–2.41)	2.06 (1.35–3.13)	<0.001
Odds ratio (95% CI)[b]	Reference	1.82 (1.28–2.59)	2.04 (1.32–3.16)	<0.001
Hyperglycemia				
Hyperglycemia cases, n(%)	88 (12.2)	43 (18.6)	26 (22.2)	
Odds ratio (95% CI)[a]	Reference	1.65 (1.11–2.46)	2.06 (1.26–3.36)	0.001
Odds ratio (95% CI)[b]	Reference	1.48 (0.97–2.27)	1.80 (1.08–3.00)	0.011

[a] Unadjusted odds ratio; [b] Estimates adjusted for age, gender, and social class

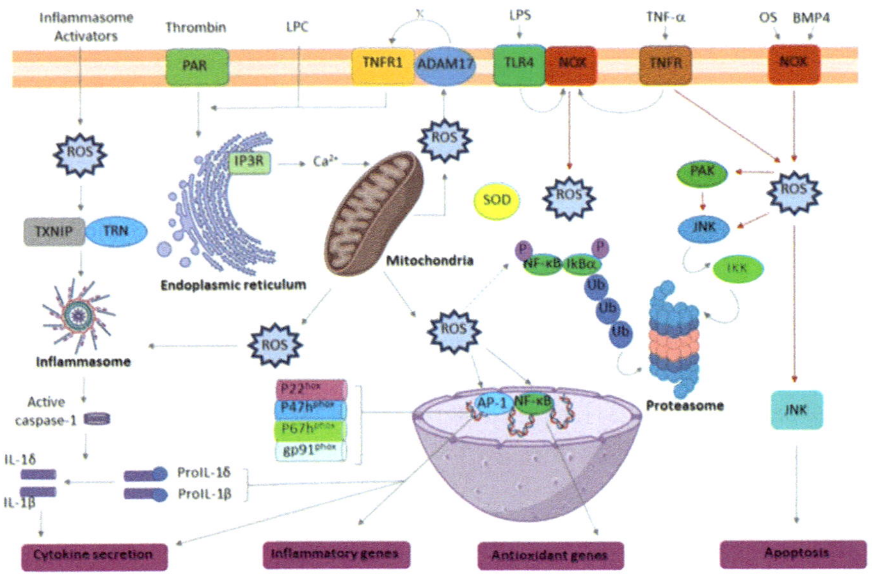

Fig. 2.9 Schematic representation of the action of reactive oxygen species (ROS) leading to inflammation. ADAM17 (ADAM metallopeptidase domain 17); ASC (Activating signal co-integrator 1); BMP4 (Bone morphogenetic protein 4); IKB-α (Inhibitor of nuclear factor kappa-B kinase regulatory subunit alpha); IKK (Inhibitor of nuclear factor kappa-B kinase); IP3R (Inositol 1,4,5-trisphosphate receptor type 3); JNK (c-Jun N-terminal kinase); LPC (Lysophosphatidylcholine); LPS (Lipopolysaccharide); NF-κB (Nuclear factor kappa subunit B); NLRP3 (NLR family pyrin domain containing 3); NOX (NADPH oxidase); OxPL (Oxidized phospholipids); PAR (Par family cell polarity regulator); PAK (p21 (RAC1) activated kinase); SOD (Superoxide dismutase); TLR4 (Toll-like receptor 4); TNF-α (Tumor necrosis factor alpha); TNFR (TNF receptor superfamily); TNFR1 (TNF receptor superfamily 1); TXNIP (Thioredoxin interacting protein); Ub (Ubiquitin) (Senevirathna 2023). Reproduced with permission through open access publishing

extract was also suggested to be due to the molecular mechanism by decreasing the production of pro-inflammatory cytokines of IL-6, IL-1, and the expression of COX-2, and simultaneous elevation in the levels of anti-inflammatory cytokine IL-4 (Bhandare 2011). Procyanidins present in areca nut extracts are reported to have antioxidant and anti-inflammatory properties. In a study by Huang et al., various procyanidins (Fig. 2.19) were extracted from areca nuts, and their potential anti-inflammatory properties were elucidated. Based on their study, procyanidins in areca nut were classified into catechins based on procyanidins having 8.7 polymerization units. The anti-inflammatory potential of the procyanidins increased with the degree of polymerization, and 12-O-tetradecanoylphorbol-13-acetate (TPA) induced inhibition of cyclooxygenase-2(COX-2) protein expression was suggested to be the main reason for the anti-inflammatory activity (Huang 2010). Further, the arecanut procyanidins also have the potential to inhibit gluconeogenesis and reduce

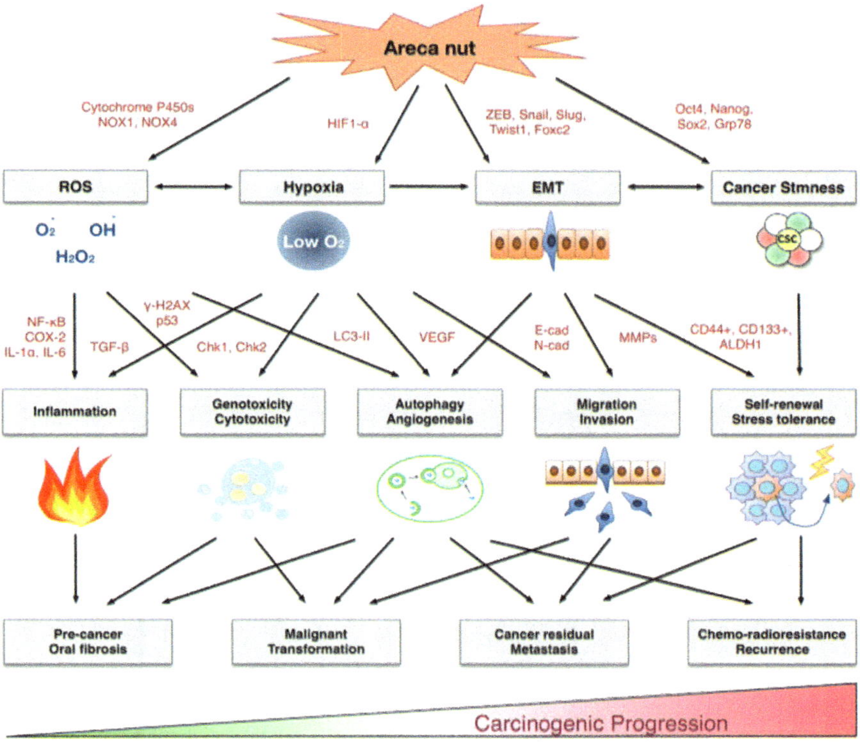

Fig. 2.10 Multifaceted mechanisms of areca nuts in oral carcinogenesis: the molecular pathology from precancerous lesions to malignant transformation. ROS: Reactive oxygen species. EMT: Epithelial-mesenchymal transition (Li 2019). Reproduced with permission through Creative Commons Attribution License

mRNA expression in phosphoenol-pyruvate carboxykinase (PEPCK) and glucose-6-phosphatase (G6Pase), which helps in increasing hyperglycemia and hence beneficial for treating diabetes (Huang 2013). Areca procyanidins are also found to cause pro-apoptotic effects in primary lymphocytes, leading to immunomodulatory effects. The extent of influence was found to be dependent on the length of the oligomer chain, with five degrees of polymerization being the most effective. Reduction in the cellular thiols by procyanidins was also considered to lead to apoptosis (Wang 2009).

Components in areca nut extracted using ethanol and methanol showed significant antibacterial activity, with methanolic extracts inhibiting both gram-positive and gram-negative bacteria with minimum inhibitory concentration of 1.56 mg/mL. Highest activity could be obtained when 100 mg/mL of the extract was used in baked cakes without adversely affecting sensory features, including texture, flavor, and smell (Table 2.17) (Jam 2021; Rahman 2014). In another study, it has been reported that the hydroalcoholic extracts (14.4% yield) from areca nuts, having total

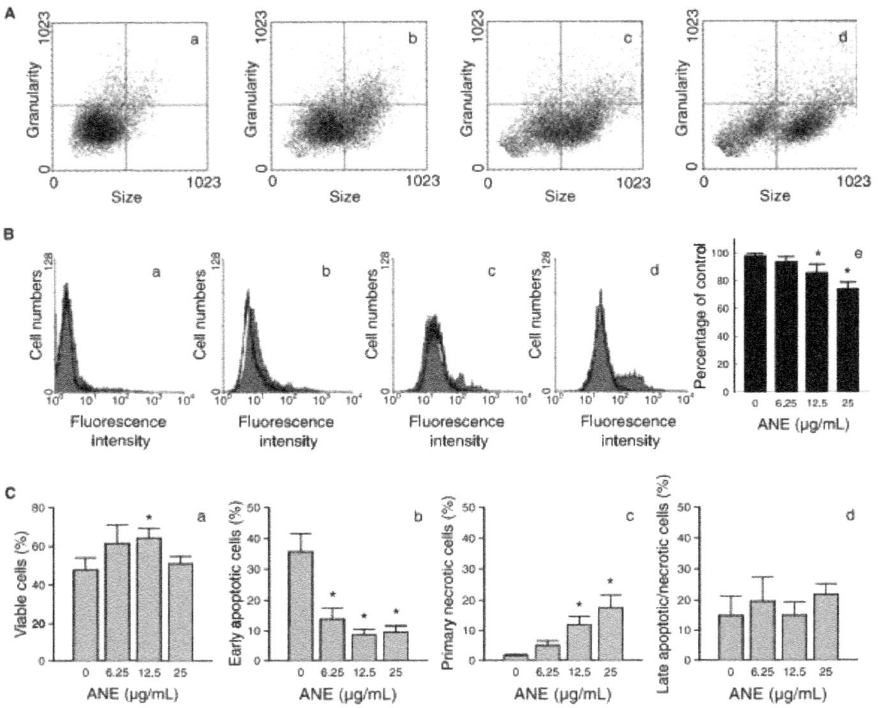

Fig. 2.11 Effect of areca nut extract on the granularity, cell viability, and apotopsis (Ho 2010). Reproduced with permission through John Wiley and Sons

phenolic content of 121 mg QE/g showed considerable analgesic, anti-inflammatory, and antioxidant potential with IC_{50} values of 83.14 μg/ml (Bhandare 2010). Results obtained using areca nut extracts were better than chemical analgesic agents such as pentazocine and aspirin. Areca nut extracts obtained using 90% ethanol contained alkaloid, tannin, and flavonoid components, with catechin and quercetin being the major constituents (Fig. 2.20). Administering this extract into mice infected with *S. aureus* led to enhanced production of white blood cells without affecting red blood cells, hemoglobin, and hematocrit levels. A dosage level of 1000 mg/kg of body weight was found to be ideal without any adverse effects on the liver and kidney (Sari 2020). Ethanolic extract of areca nut has also shown to have antidepressant activity in both acute and sub-chronic forced swim tests due to the presence of saponins, whereas alcoholic extracts from arecanuts showed toxic effects on rats due to the production of serotonin and nonadrenaline (Abbas 2013). Similarly, ethanolic areca nut extracts showed significant reduction in mounting latency (ML) and intromission latency (IL) in Swiss albino mice, which was attributed to the aphrodisiac effect of the areca extract (Anthikat 2013).

In another study, it has been demonstrated that water-soluble areca extracts are capable of inhibiting cholesterol absorption in high cholesterol-fed rats particularly

Table 2.10 Effects of tobacco and areca nut consumption on female reproduction in animals and humans (Kumar 2013)

Sl. no	Products/toxicants	Observed effects
Animal study		
1	The antiovulatory and abortifacient effects of areca catechu in female rats	The ethanolic extract of Areca catechu at doses of 100 and 300 mg/kg has antiovulatory and abortifacient effects in rats
2	The effect of arecoline—major alkaloid of areca nut on implanted embryos in early pregnant mice	Arecoline decreased the number of implanted embryos in ear pregnant mice. In addition, trophoblast outgrowth expansion of blastocysts was also inhibited. Suggest that arecoline is toxic to mouse embryos as early as peri-implantation
3	Determined arecoline in biological matrices accounting fetal exposure, exposure associated with placental abnormalities, and neonatal withdrawal syndrome	The recognized biochemical effects of arecoline on the autonomic nervous system and the embryo toxicity in animals, together with observed findings, allowed to hypothesize a number of effects on the fetus related to abnormalities of the feto-placental circulation, similar to those observed with nicotine or cocaine
Human study		
1	Maternal smokeless tobacco use, mostly oral applications of burnt tobacco or 'mishri', in pregnancy	Maternal use of 'mishri' tobacco in pregnancy may be associated (1) the offsprings' low birth weight, (2) low birth weights in girls more than in boys, (3) decreased male: female ratio, (4) low birth weight of <2.0 kg more than 2–2.5 kg
2	Maternal use of Mishri (Tobacco containing teeth cleaning powder) on the fetus and the perinatal outcome	The complications during the previous perinatal period, current pregnancy/delivery and stillbirths were significantly more among Mishri users. A relative risk of abnormal delivery was 2.7 for users. 153 Mishri user women gave birth to babies weighing on an average 169.9 g less (significant) than babies of mother who never used
3	500 women enrolled, 167 consumed tobacco during pregnancy	In women who used no tobacco, the LBW was 36.28%, and in those consumed tobacco in pregnancy the incidence was 64.62% indicating role of tobacco consumption in LBWs
4	Smokeless tobacco use during pregnancy	Increases stillbirth risk, with a risk that as is associated with maternal cigarette smoking
5	Effects of maternal prenatal smokeless tobacco use on infant birth size	The mean birthweight of infants of smokeless tobacco users was reduced by 78 g (non-significant) compared non-users and by 331 g in infants of smokers. No association found between maternal smokeless tobacco use and infant length or head circumference

(continued)

Table 2.10 (continued)

Sl. no	Products/toxicants	Observed effects
6	A total of 150 mothers using ST for 5 years or more and their recently delivered newborn infants (mother-neonate pair) and 150 matched mother-neonate pairs as cases and controls respectively	Use of ST for >5 years was significantly associated with spontaneous abortion with a 2.3 times more risk than that of non-ST users. It was also significantly associated with stillbirth $P < 0.01$) with a risk of stillbirth two times higher than that of non-ST users. Use of ST for >5 years was significantly associated with preterm delivery ($P < 0.001$) and risk 3.1 times more than that of non-ST users. It was also significantly associated with delivery of LBW baby ($P < 0.001$) with a risk of LBW baby 4.1 times higher than that of non-ST users
7	Risk of betel quid use on birth outcomes, a total of229 women recruited, 32 women with adverse birth outcomes constituted the case group	Adverse birth outcomes were associated with maternal betel quid chewing and maternal age. After adjusting for age, the risk of adverse birth outcome was five times higher among betel quid chewing women
8	Betel nut chewing and impact on pregnancy. Among pregnant women, 94% chew betel nut, 9% smoke and 1% used alcohol	Primigravidity, betel nut chewing and low BMI had a statistically significant impact on birth weight reduction of 467 g, 238 g and 175 g respectively

Reproduced with permission from Elsevier

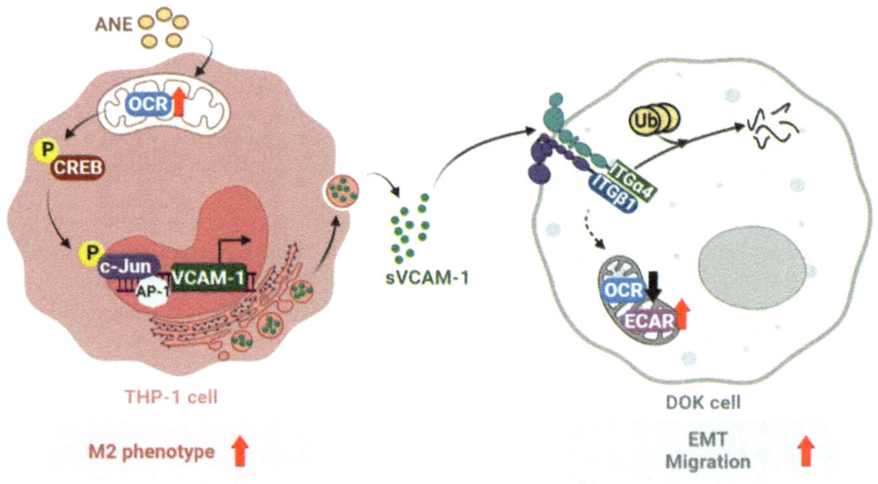

Fig. 2.12 Schematic drawing for malignant transformation in oral premalignant cells via ANE-induced M2 macrophage differentiation (Yuan 2024). Reproduced from Springer Nature with permission through open access publishing

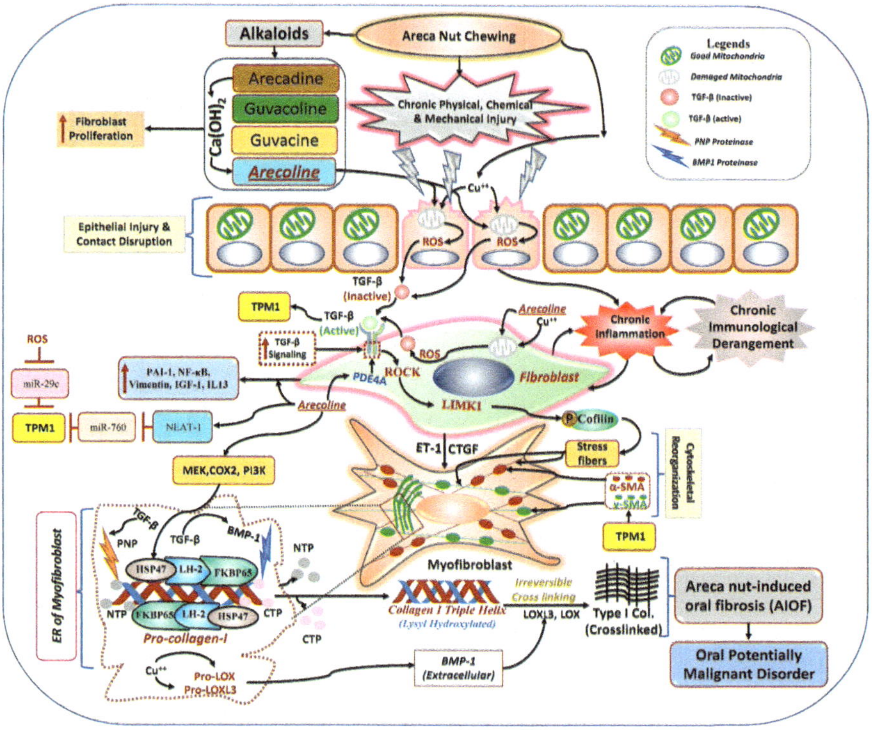

Fig. 2.13 Oral submucous fibrosis with a particular proclivity for malignant transformation is re-categorized as—Areca nut-induced oral fibrosis (AIOF) (Sharma 2024). Reproduced from Elsevier with permission through open access publishing

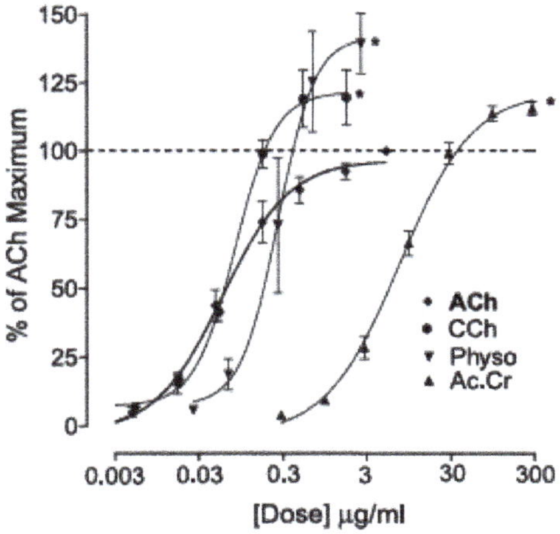

Fig. 2.14 Comparative dose–response curves of physostigmine (Physo), carbachol (CCh), acetylcholine (ACh), and Areca catechu crude extract (Ac.Cr) in isolated rabbit jejunum (Gilani 2004). Reproduced with permission from Elsevier

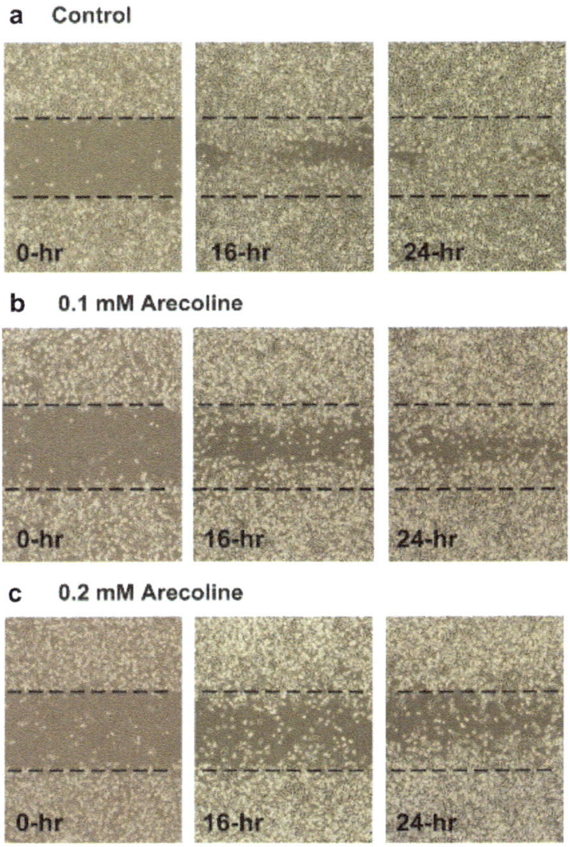

Fig. 2.15 Closure of a wound by EAHY cells photographed initially (0 h) and after 16 and 24 h of incubation with or without arecoline. **a** Control cells, **b** 0.1 mM arecoline. Compared to the control, fewer EAHY cells had migrated into the wounded area after exposure to 0.1 mM arecoline for 16 and 24 h. **c** Arecoline at 0.2 mM. The migration of EAHY cells into the wounded area was markedly inhibited. Dotted lines (- - -) indicate the original wounded area. (40×, original magnification) (Tseng 2014). Reproduced from Elsevier with permission through open access publishing

triglyceride absorption and plasma lipid concentration. Such reduction was possible due to the lower production of small intestinal pCEase (pancreatic cholesterol esterase) activity by 22.5% and reduced activity of hepatic and intestinal ACAT (acyl-CoA:cholesterol acyltransferase) (Byun 2001). Similarly, water-soluble extracts of areca were able to significantly reduce plasma cholesterol concentration by 25% without changing the triglyceride concentration. Similarly, intestinal pCEase activity was reduced by 39.1%. Arecanut extracts were not only able to reduce intestinal enzyme activity but also lead to lowered absorbed cholesterol in blood, which could promote metabolic efficiency (Park 2002).

In a distinct study, Shin et al. have reported that arecoline could have cognitive-enhancing effects in patients with Alzheimer's dementia. When injected into mice in 50–200 μM dosage, there was neuronal cell death and increased reactive oxygen species (ROS) production and upregulation in protein expression and mRNA levels of NADPH oxidase 2. An increase in the expression of pro-apoptotic proteins

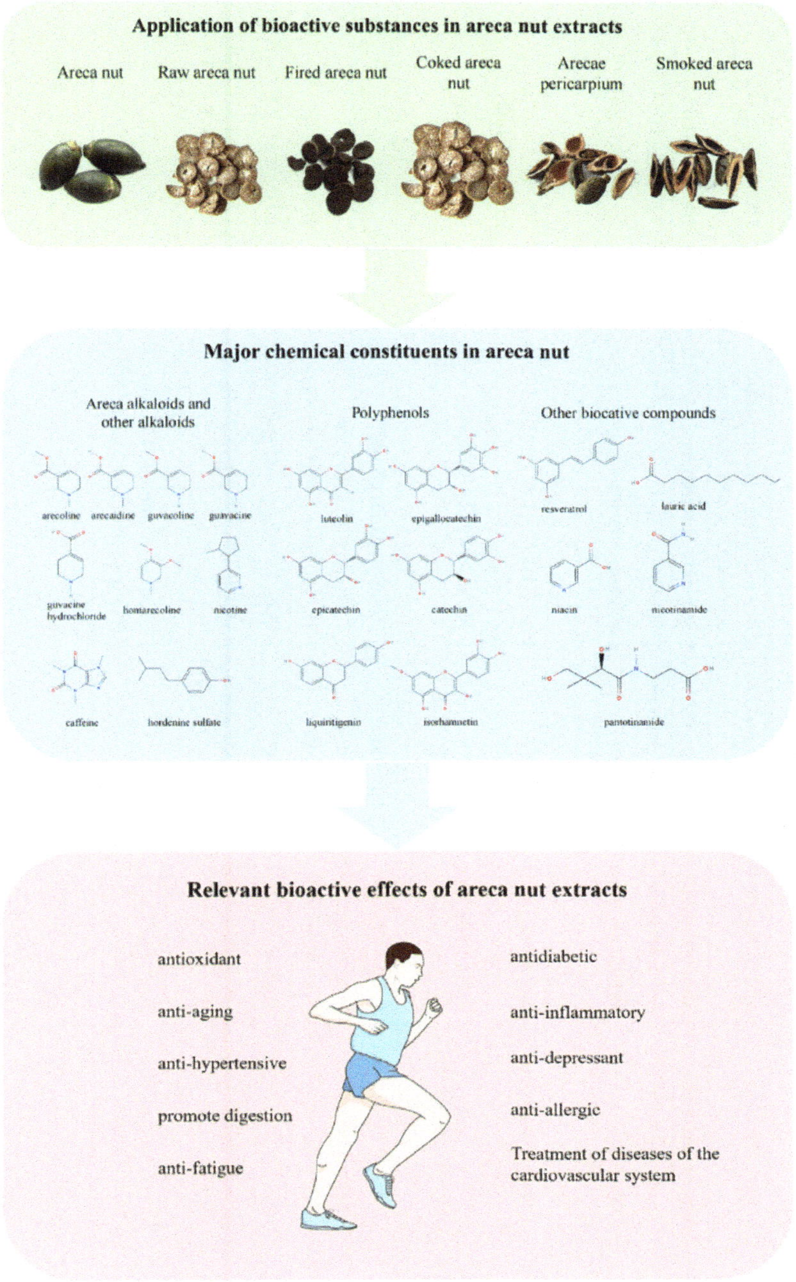

Fig. 2.16 Types, major components, and biological benefits of areca nut extracts (Guo 2024). Reproduced with permission through open access publishing

Table 2.11 Summary of the beneficial functions and mechanisms of areca nut extracts and isolated bioactive compounds (Sun 2024)

Extracts/ingredients		Functions	Model	Mechanism	Dose
Areca nut extract	Aqueous fraction	Anti-inflammatory	Mice and rats	↓ Arachidonic acid metabolism ↑ degradation and/or inactivation of PGE2	100 mg/kg (p.o.)
	Aqueous fraction	Antidepressant	Rats	↑ Secretion of serotonin	10 mg/kg (i.p.)
	Hexane fraction	Antidepressant	Mice and rats	↓ MAOs	2.5 mg/kg (i.p.)
	70% aqueous–methanol extract	Hypolipidemic	Rats	↓ pCEase and pancrelipase	10 mg/kg (i.g.)
	80% (v/v) aqueous ethanol extract	Anti-allodynic	Rats	↑ α-2 adrenoceptors and 5-HT7 receptors	300 μg (intrathecal administration)
	90% aqueous–methanol extract	Hypolipidemic	Rats	↓ ACAT and pCEase activity	0.5% (p.o.)
	Methanol extract	Antibacterial	In vitro	↑ Destroying the cell wall	
	Ethylene extract	Deworming	In vitro	↓ Carbohydrate metabolism of the parasite	
	Areca nut extract	Deworming	Chicks	↓ The causative agent Eimeria tenella	100 mg/kg (p.o.)
		Antitumor	Mice	↑ ROS, autophagy, and lysosomal formation	20 mg/kg (i.p.)
	Pericarpium arecae extract	The improvement of intestinal peristalsis	Guinea pigs	↑ Cholinergic M3 receptors	1 g/mL concentrated stock solution of crude drug (3% concentration)
	Semen Arecae and Charred Semen Arecae extract	Gastrointestinal protection	Rts	↓ Cholecystokinin mRNA expressions ↑ serum levels of substance P and motilin	3 g/kg (p.o.)

(continued)

Table 2.11 (continued)

Extracts/ingredients		Functions	Model	Mechanism	Dose
	Areca nuts powder extracted with supercritical carbon dioxide	Deworming	Chicks	↓ Eimeria tenella	100 mg/kg (p.o.)
	Ethylene extract	Deworming	In vitro	↓ Carbohydrate metabolism	0.5 mg/mL
	Crude aqueous–methanol extract	Deworming	Sheep	↓ Hatching of parasite eggs, the development of the infected larval stage, and the physiological functions of the parasite	0.33 g/kg (p.o.)
Alkaloids	Total alkaloids	Analgesia	Mice	↓ COX-2 expression	100 mg/kg (i.g.)
	Arecoline	Anxiolytic	Zebrafish	↑ Norepinephrine ↑ Serotonin	10 mg/mL
		Refreshing		↑ Parasympathetic nerves	
			Zebrafish	↑ Central M receptors	0.01 ppm
			Rats	↑ HPA axis	0.2 mg/kg (i.p.)
			Rats	↑ VTA dopaminergic neurons	0.2 mg/kg (i.v.)
		Antibacterial	In vitro	↓ *Bacillus subtilis, Enterococcus faecalis, Escherichia coli, Pseudomonas aeruginosa, Staphylococcus aureus, Streptococcus*	
		Relieve schizophrenia	Mice	↓ Cerebral demyelination	2.5 mg/kg/day (p.o.)
		Relieve Alzheimer's disease	Rats	↑ Cerebral blood flow and metabolism	2 mg/kg (p.o.)
		Hormone levels regulation	Mice	↑ Muscarinic cholinergic receptors	10 mg/kg (i.p.)

(continued)

Table 2.11 (continued)

Extracts/ingredients		Functions	Model	Mechanism	Dose
		Antitumor	BCC-1/KMC and HaCaT	↓ IL-6 ↑ P53 and apoptosis	30 μg/mL
			In vitro	Alters the cell cycle ↓ cell viability ↓ cancerous prostate cells	0.4 mM
		Gastrointestinal protection	Grass carp	↑ M3 ↑ Keap1a/Nrf2 signaling pathway ↓ RhoA/ROCK signaling pathway	1 mg/kg (p.o.)
		Anti-osteoporotic	Mice	↑ Osteoblast differentiation ↓ osteoclastogenesis	5 mg/kg (p.o.)
	Arecaidine	Refreshing	In vitro	↓ GABA uptake	
	Guvacine				
Phenols	Total phenols	Hypoglycemic	In vitro	↓ α-glucosidase activity	IC_{50} 1.50 ± 0.31 μg/mL
		Anti-osteoporotic	Rats	↓ 5-HT ↑ bone resorption and formation	400 mg/kg (i.g.)
			In vitro	↑ Proliferation, differentiation, and mineralization of osteoblasts	25 μg/mL
		Anti-fatigue	Mice	↑ Lactate dehydrogenase, superoxide dismutase, and catalase activities ↓ malondialdehyde content	20 mg/kg (i.g.)
		Antioxidant	RAW264.7 cells	↓ MAPK pathway ↑ Nrf2/HO-1 antioxidant pathways	40 μg/mL
		Anti-viral	In vitro	↓ HIV type 1 protease	0.2 mg/mL

(continued)

Table 2.11 (continued)

Extracts/ingredients		Functions	Model	Mechanism	Dose
	Procyanidins	Analgesia	Mice and rats	↓ Plasma protein extravasation and inflammation in the dura	250 mg/kg (p.o.)
		Gastrointestinal protection	Mice	↓ Th2 responses ↑ induction function of myeloid-derived suppressor cells	0.05% in water
		Hypoglycemic	Mice	↓ G6Pase and PEPCK	10 mg/kg/day (i.g.)
		Anti-photoaging	Mice	↓ Cyclooxygenase-2, matrix metalloproteinase, and oxidative stress	10 mg/kg (p.o.)
		Anti-hypertensive	In vitro	↓ Angiotensin-converting enzyme	IC50 1.51 ± 0.65 mg/mL
		Anti-viral	In vitro	↓ HIV type 1 protease	0.2 mg/mL
	Isorhamnetin	Antidepressant	Mice	↓ MAOs	0.2 mL/10 g (i.g., concentration of 2.5 mg/mL)
	Chrysoeriol Luteolin Chrysophanol				
	Syringic acid	Antioxidant	In vitro	↓ DPPH radical scavenging activity, hydroxyl radical scavenging activity, and reducing power	EC50 0.409 mg/mL and 0.188 mg/mL, respectively
	Epicatechin				
		Anti-hypertensive	In vitro	↓ Angiotensin-converting enzyme	IC_{50} 1.51 ± 0.65 mg/mL
	Catechin				
	Arecatannin	Anti-viral	In vitro	↓ HIV type 1 protease	0.2 mg/mL
Triterpenes and Steroids	Fernenol	Hypoglycemic	In vitro	↓ α-glucosidase activity	IC_{50} 1.50 ± 0.31 μg/mL
	Arundoin				EC50 47.5 mg/L
	The mixture of stigmasterol and β-sitosterol				EC50 56.7 mg/L

(continued)

Table 2.11 (continued)

Extracts/ingredients	Functions	Model	Mechanism	Dose
Polysaccharide	Anti-inflammation	Raw264.7 cell	↓ NO production	IC_{50} 85.64 mg/mL
	Antioxidant	In vitro	↓ DPPH/hydroxyl radicals	2 mg/mL
Fatty acids	Antibacterial	In vitro	↓ Bacillus subtilis, Enterococcus faecalis, Escherichia coli, Pseudomonas aeruginosa, Staphylococcus aureus, Streptococcus	
Areca fruit	Relieve epilepsy	Human	↑ Muscarinic receptor	Three nuts/day

↑: Activated. ↓: Inhibited. p.o.: Oral administration. i.g.: Oral gavage. i.p.: Intraperitoneal injection. i.v.: Intravenous injection. EC_{50}: Half maximum effective concentration. IC_{50}: Half maximal inhibitory concentration.

Reproduced with permission through open access publishing

Table 2.12 Effect of substrate concentration and extraction time on the components extracted from areca nuts (Chavan 2013)

Arecanut powder conc. Wt, %	Total phenolic content (TPC) (mg GAE/g)	Total tannin content (TTC) (mgCE/g)	Arecoline (mg/g)	TPC, arecoline	ABTS, (mmol/L TEAC)	FRAP (mmol/L TEAC)	Hydroxyl scavenging activity, %
6	217 ± 1.30	66 ± 0.83	1.14 ± 0.01	190	225 ± 2.05	383 ± 1.30	27 ± 0.11
8	256 ± 0.01	71 ± 0.74	1.24 ± 0.01	206	263 ± 1.60	590 ± 1.02	31 ± 0.80
10	320 ± 1.01	79 ± 1.07	1.34 ± 0.02	239	459 ± 1.20	1011 ± 1.11	53 ± 0.25
12	381 ± 1.67	102 ± 0.20	2.09 ± 0.02	182	478 ± 1.03	1042 ± 1.21	60 ± 0.46
14	396 ± 1.25	105 ± 0.48	2.58 ± 0.01	154	502 ± 1.20	1062 ± 1.14	62 ± 0.02

Reproduced with permission from John Wiley and Sons

(cytochrome *c*, Bax, caspase-9, and caspase-3) and increased expression of anti-apoptotic protein Bcl-2 were observed which could cause neuronal apoptopic death (Shih 2010).

Not only humans, even animals have also profited from the beneficial properties of areca nut extracts. Feeding 270-day-old broiler chicks with areca nut extract significantly improved feed intake and body weight. Further, the extracts increased concentration of interleukin 2 (IL-2), which provided anticoccodial properties and beneficial effects on intestinal mucosa damage in affected chicks (Wang 2018). In another study, polyphenols were extracted from areca nut using an ultrasonic cleaner. The yield of the polyphenols was dependent on the temperature, time, ethanol concentration, and ultrasonic power (Fig. 2.21). The maximum yield of phenols was 139 mg/g, and the extracted components were able to induce proliferation, differentiation, and mineralization of MC3T3-E1 osteoblasts (Sun 2023). Nearly 98 volatile compounds (Table 2.18) could be extracted from areca nuts using headspace solid-phase microextraction (HS-SPME, DVB/CAR/PDMS fiber extraction) and simultaneous hydrodistillation extraction (SHDE). Fatty acids were the major components extracted, which showed good antimicrobial activity and ability to be used as natural antibiotic (Machova 2023). Ability of areca nut extracts to reduce burn wound model in rats was studied in comparison to dexamethasone-treated rats. Areca nut extracts were combined with gum acacia and made into an oral formulation for rats with burn wounds. Areca nut-treated rats showed significantly higher wound contraction rate and period of epithelialization as seen from Tables 2.19 and 2.20 (Bharat 2014).

Table 2.13 Effect of extraction time on arecanut constituent (Chavan 2013)

Extraction time, min	Total phenolic content (TPC) (mg GAE/g)	Total tannin content (TTC) (mgCE/g)	Arecoline (mg/g)	TPC, arecoline	ABTS, (mmol/L TEAC)	FRAP (mmol/L TEAC)	Hydroxyl Scavenging activity, %
30	212 ± 2.38	56 ± 1.66	0.86 ± 0.01	247	217 ± 1.13	689 ± 1.40	14.9 ± 0.13
60	399 ± 2.70	97 ± 0.77	1.24 ± 0.02	322	481 ± 1.08	1038 ± 1.51	552.4 ± 1.10
90	407 ± 1.31	102 ± 2.82	1.73 ± 0.03	236	505 ± 1.20	1043 ± 0.83	61.4 ± 1.09
120	392 ± 1.37	103 ± 1.92	2.46 ± 0.01	159	503 ± 1.23	1031 ± 1.23	52.9 ± 0.46
150	382 ± 1.37	100 ± 0.50	2.66 ± 0.01	143	476 ± 1.12	1003 ± 1.21	41.5 ± 0.30

Reproduced with permission from John Wiley and Sons

Table 2.14 Some of the chemical compounds isolated from areca nuts (Tong 2024)

Classification	Chemical component
Alkaloids	Arecoline, Arecaidine, Guavacoline, Guavacine, Arecolidine, Ethyl N-methyl-1,2,5,6-tetrahydro-pyridine-3-carboxylate, Methyl nicotinate, Ethyl nicotinate, Methyl N-methylpiperidine-3-carboxylate, Ethyl N-methylpiperidine-3-carboxylate, Nicotine, Isoguvacine, Homoarecoline
Flavonoids	Isorhamnetin, Chrysoeriol, Luteolin, Quercetin, 4′,5′-dihydroxy-3′,5′,7′-trimethoxyflavonone, 5,7,4′-trihydroxy-3′,5′-di methoxy flavanone, Liquiritigenin, Jacareubin,
Tannins	Catechin, Epicatechin, Procyanidin A1, Procyanidin B1, Procyanidin B2, Arecatannin A1, Arecatannin B1, Arecatannin C1, Arecatannin A2, Arecatannin A3, Arecatannin B2
Triterpenoids and steroidals	Arborinol, Arborinol methyl ether, Ursonic acid, 3β-acetyl ursolic acid, Cycloartenol, Fernenol, Arundoin, 5,8-epidioxiergosta-6,22-dien-3β-ol, Stigmasta-4-en-3-one, β-sitosterol
Fatty acids	Lauric acid, Myristic acid, Palmitic acid, Stearic acid, Oleic acid
Others	Chrysophanol, Physcion, Resveratrol, Ferulic acid, p-hydroxybenzoic acid, Vanillic acid, de-O-methyllasiodiplodin, Epoxyconiferyl alcohol, 4-[3′-(hydroxymethyl) oxiran-2′-yl]-2,6-dimethoxyphenol, Protocatechuic acid, Isovanillic acid, Cyclo-(Leu-Tyr)

Reproduced with permission through open access publishing

2.4 Novel Uses of Areca Nuts

2.4.1 Synthesis of Nanoparticles

2.4.1.1 Silver Nanoparticles Synthesized Using Areca Nut Extracts

Presence of alkaloids and reducing agents in areca nut extracts has been considered useful for the synthesis of nanoparticles. Bhat et al. treated areca nuts in distilled water under microwave for 180 s. The extract obtained was filtered and used as the reducing agent to synthesize silver nanoparticles from silver nitrate solution (Bhat 2013) through microwave exposure for 10–90 s. Nanoparticles of different sizes were obtained depending on the concentration of areca nut extract used and the treatment time in the microwave. As seen from Fig. 2.22, nanoparticles with size between 7.32 and 15.91 nm were obtained. In a similar study, aqueous extracts from areca nuts were used as both capping and reducing agent to rapidly synthesize silver nanoparticles within 90 s. Presence of epicatechin, rutin, procyanidin B1, gallic acid, arecolidine, and arecoline was suggested to be responsible for the reduction of silver cations from the silver nitrate solution. Size of the nanoparticles was dependent on the conditions prevailing during synthesis, but acid conditions resulted in larger particles compared to alkaline conditions. Synthesized nanoparticles were stable up to 75 days and exhibited antioxidant properties (Habeeba 2024). Areca nut extract could act as both a reducing agent and capping agent during the synthesis of nanoparticles from silver

Table 2.15 Major pharmacological and associated properties of areca nut and their extracts (Tong 2024)

Pharmacological effects	Main bioactive phytoconstituents	Main experimental methods	Dosage	Main mechanism of action
Antidepressant activity	Ethanol, hexane, ethyl acetate, aqueous extract fractions of the AN	Forced swim test; tail-suspension test; locomotion test; Monoamine oxidase assay	2.5, 5, 10, 13, 20, and 50 mg/kg	The alkaloids viz arecaine, arecaidine, and arecoline have been shown to be present in the AN, with only arecoline having MAO-inhibiting properties. The possibility that the MAO inhibition in the present report may be due to arecoline and some other unknown compounds needs further investigation
	Dichloromethane and aqueous extract of the AN	Tail-suspension test; yohimbine potentiation test; locomotor test; monoamine oxidase assay	5 and 10 mL/kg for rats and mice	Inhibiting monoamine oxidase type A in rat brain homogenates
	Ethanol extracts from AN	Acute and sub-chronic forced swimming tests	AN ethanol extract (50 mg/kg), aqueous extract (20 mg/kg)	Increased serotonin levels by approximately 35% and norepinephrine levels by approximately 30%
	Areca Thirteen Pill (GY-13)	Sucrose preference tests, open field tests, and body weight measurements	0.25 g/kg (low), 0.50 g/kg (medium), 1.00 g/kg (high)	Increases the levels of cAMP and PKA, restores the mRNA levels of CREB and BDNF, and enhances growth activity in the hippocampus
Anti-inflammatory activity	Acetone, ethanol extract of ANs	Paw edema in rats; free radical scavenging activity	10, 50, and 100 mg/kg	Carrageenan in rats and reduce the levels of prostaglandin E2; the inhibition of cyclooxygenase
	Active components of ANs	Quantitative real-time PCR analysis; detection of IL-6 production; apoptosis analysis; cell cycle analysis	2 mg/kg	Lower the levels of interleukin-6 (IL-6); increase the levels of the tumor suppressor factor p53

(continued)

Table 2.15 (continued)

Pharmacological effects	Main bioactive phytoconstituents	Main experimental methods	Dosage	Main mechanism of action
	The water extract derived from the pericarpium of AN	Biochemical analysis; Masson's trichrome staining Immunofluorescence staining; real-time RT-PCR; Western blot	50, 100, or 200 mg/kg	Deactivating pancreatic stellate cells; the water extract from ARP exhibited the ability to impede the constituents of the extracellular matrix (ECM), specifically targeting alpha-smooth muscle actin (α-SMA), collagen I, and fibronectin 1 (FN1) in both pancreatic tissue and pancreatic stellate cells (PSCs)
Anti-inflammatory activity	Gwakhyangjeonggi-san (GHJGS)	LPS-induced phosphorylation of mitogen-activated protein kinases; Western blotting and enzyme-linked immunosorbent assay (ELISA)	0, 250, 500, or 1000 μg/mL	Increased expression of heme oxygenase-1 and prevention of reactive oxygen species generation
Anti-inflammatory activity	Arecoline	Bone marrow macrophage (BMM) isolation; cell viability assay; osteoclast differentiation; tartrate-resistant acid phosphatase (TRAP) staining; real-time RT-PCR; Western blot	10 mg/kg	The interference of arecoline with signal pathways activated by AKT, MAPK, and NF-κB, leading to the suppression of gene expression and the translocation of genes associated with osteoclast differentiation

(continued)

Table 2.15 (continued)

Pharmacological effects	Main bioactive phytoconstituents	Main experimental methods	Dosage	Main mechanism of action
	AN polyphenols	Cell viability assays; *specific marker protein* detection; fluorescence investigations	0, 25, 50, 100, 150 μg/mL	Areca nut seed polyphenol (ACP) remarkably enhanced the content of ALP in osteoblasts, which promoted the differentiation of osteoblasts
Antioxidant activity	AN polyphenol; ethanol extract from AN	Quantitative real-time PCR analysis; Western blot; free radical scavenging activity assays	Oral administration of ANE (250 mg/kg and 500 mg/kg)	Inhibit vascularization in plasma protein-nitroglycerin venous and local models, as well as inhibit the expression of inducible nitric oxide
	AN polyphenol	Quantitative real-time PCR analysis; Western blot; free radical scavenging activity assays	IC_{50} values were 26.9 mg/mL for porcine pancreatic elastase and 60.8 mg/mL for human neutrophil elastase	The elimination of free radicals and the enhancement of antioxidant enzyme activities, such as catalase (CAT), superoxide dismutase (SOD), and glutathione peroxidase (GPX)
	AN polyphenol	Free radical scavenging activity assays	IC_{50} = 1483.43 μg DW/mL	Inhibiting elastase, reduction in skin tissue aging and inflammation
	EtOAc extract from AN	DPPH radical scavenging assay; hydroxyl radical scavenging activity	20, 40, 60, 80, 120, and 200 μg/mL	The polyphenols in the ethanol extract of areca nut seed can remove free radicals and exert antioxidant activity

(continued)

Table 2.15 (continued)

Pharmacological effects	Main bioactive phytoconstituents	Main experimental methods	Dosage	Main mechanism of action
Antimicrobial activity	Methanol, ethanol, and water extract from ANs	Agar disk diffusion assay; microtiter broth dilution method; agar disc diffusion technique	The agar disc diffusion technique: 100 mg/L	AN is rich in tannins, terpenoids, alkaloids, and flavonoids, and tannin in AN is effective to inhibit gram positive bacteria. Tannin binds with peptide on the peptidoglycan component from cell walls, which in turn disturbs the integrity of bacterial cell walls, which cause bacterial cell damage. Finally, it led to disturbance of the metabolism process and, subsequently, the death of cells
	AN polyphenols	Microwave and Soxlet apparatus; resazurin microtiter assay	MIC: 0.975 ± 0.02 μg/mL	The bioactivity of the extract was attributed to the nontoxic polyphenols present. This extract also showed selective inhibition of M. tuberculosis over other gram-positive and gram-negative bacteria

Reproduced with permission through open access publishing

Table 2.16 Summary of the systemic effects of areca nut (Garg 2014)

Site	Effect of Areca nut	Pathway/mediators
Nervous system	• Euphoria and dependency syndrome • Increased skin temperature Salivation • Palpitation • Heightened alertness • Antimigraine effect Neurotoxicity	• GABA inhibition by Arecoline • Arecoline affects the ANS • Increased α and β activity • Inhibition of iNOS • Suppression of anti-oxidative mechanisms
Cardiovascular system	• Tachycardia and increased systolic BP • Decreased diastolic BP • Increased atherogenesis • Coronary artery spasm • Increased risk of CAD	• Central sympathetic response • Peripheral cholinergic effect • Blocking of HDL receptors and inhibition of LDL uptake by the liver • Parasympathomimetic effect of arecoline • Artherogenisis and coronary artery spasm
Gastrointestinal system	• Type II DM, hyperlipidemia, hypertriglyeridemia and metabolic syndrome • Hepatotoxicity • Laxative and sialogogue effect • Decreased growth in weight and BMI	• Arecoline inhibits adipocyte differentiation, and insulin promotes glucose uptake • G0-G1 cell cycle arrest and DNA damage • Action of arecoline on M3 receptors and AChE action
Endocrine system and reproductive health	• Thyroid • Acute effect increased T3, T4 and decreased TSH • Chronic effect-hypothyroidism • Prostate hyperplasia Infertility • Fetus – Low birth weight and length, preterm labor Increased exposure to teratogens and carcinogens Arrest of endothelial cell differentiation Vitamin D deficiency	– Arecoline – Increased expression of androgen receptors – Changes in sperm count, motility, and morphology – Arecoline – Increased expression of 25(OH)ase
Blood and its components	• Increased fibrogenesis Decreased production of IL-2 and IFN γ • Suppresses T cells and cytokine Th1 • Cytotoxic to RBCs	• Increased platelet aggregation, Ca++, and TXA release • Cytotoxic to splenocytes • Oxidative stress • Damages the RBC membrane

(continued)

Table 2.16 (continued)

Site	Effect of Areca nut	Pathway/mediators
Leukotrines and arachidonic pathway	• Analgesic at high doses Antiinflammatory • Carcinogenic and mutagenic effect	• Central and peripheral pathways • Decreased release of leukotrienes, prostaglandins, and histamines • Stimulation of gingival keratinocytes • Increased peroxyl radicals • Increased malonaldehyde
Respiratory system Acute toxicity of areca nut	– Aggravation of asthma Decreased FEV1 – Dyspnea, tachycardia, palpitations, chest tightness, vomiting, vertigo, dizziness, abdominal colic, rarely myocardial infarction, and coma	• Arecoline • Arecoline and other alkaloids

Reproduced with permission from Georg Thieme Verlag KG

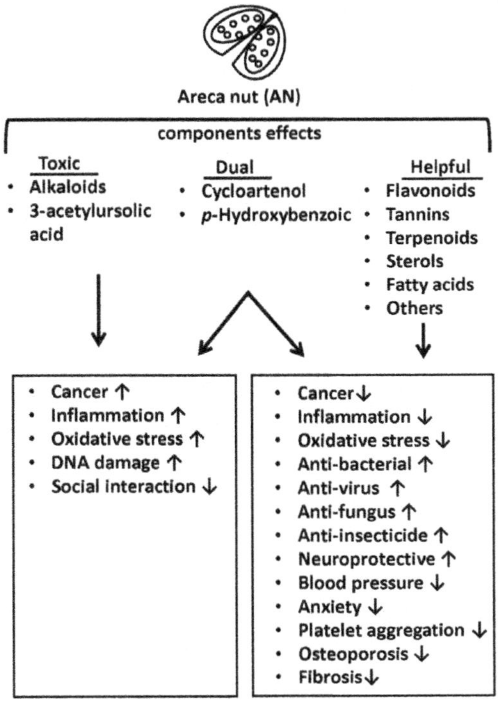

Fig. 2.17 Pharmacological and toxicological effects of AN components. Most of the AN components are helpful for organisms from a health point of view. However, some AN components have toxicological effects. ↑: Activated; ↓: Inhibited (Liu 2023). Reproduced with permission through open access publishing

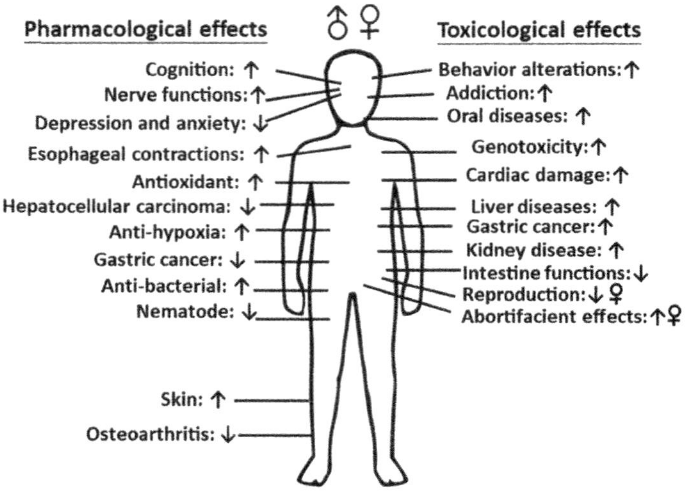

Fig. 2.18 The distribution of pharmacological and toxicological effects of AN extract, combined compounds, and arecoline in the human body. The effects in other animals, such as mice, rats, porcine, zebrafish, C. elegans, chickens, and buffaloes were included from different sources cited in this review. ↑: Activated, ↓: Inhibited (Liu 2023). Reproduced with permission through open access publishing

nitrate using microwave approach. Particles obtained had average size of 17 nm and were composed mostly of silver with minor amounts of carbon and nitrogen (Fig. 2.23) (Dash 2022). Nanoparticles showed excellent antibacterial properties and were also non-cytotoxic to human blood cells.

Using a similar approach, silver nanoparticles having size of about 10 nm were synthesized using pH 12 when silver nitrate concentration of 1 mM were employed as the optimum concentration. The nanoparticles (spherical shaped) had a zeta potential of -30.50 mV and polydispersity of 0.328 (Fig. 2.24). Nanoparticles also possessed excellent free radical scavenging potential with IC_{50} value of 44.85 μg/mL, significantly higher than the common antioxidant chemical ascorbic acid. Likewise, nanoparticles also showed high activity against six common pathogens (Hu 2022). Aqueous extract from areca nuts was used to prepare silver nanoparticles with high antioxidant activity and potential application in catalysis. Spherical and crystalline nanoparticles with diameter between 18.2 and 24.3 nm were obtained. The nanoparticles were able to degrade organic pollutants such as methylene blue and 4-nitrophenol and were also able to scavenge free radicals suggesting suitability for environmental and biomedical applications (Rajan 2015). Silver nanoparticles (45 nm diameter) synthesized using aqueous areca nut extracts were studied for their electrolytic activity and potential use as glucose detector. Cyclic voltammetry studies showed excellent electrocatalytic activity with peak current density of 4.14×10^{-4} A, higher than that of bare glass electrode (Fig. 2.25) (Edison 2017).

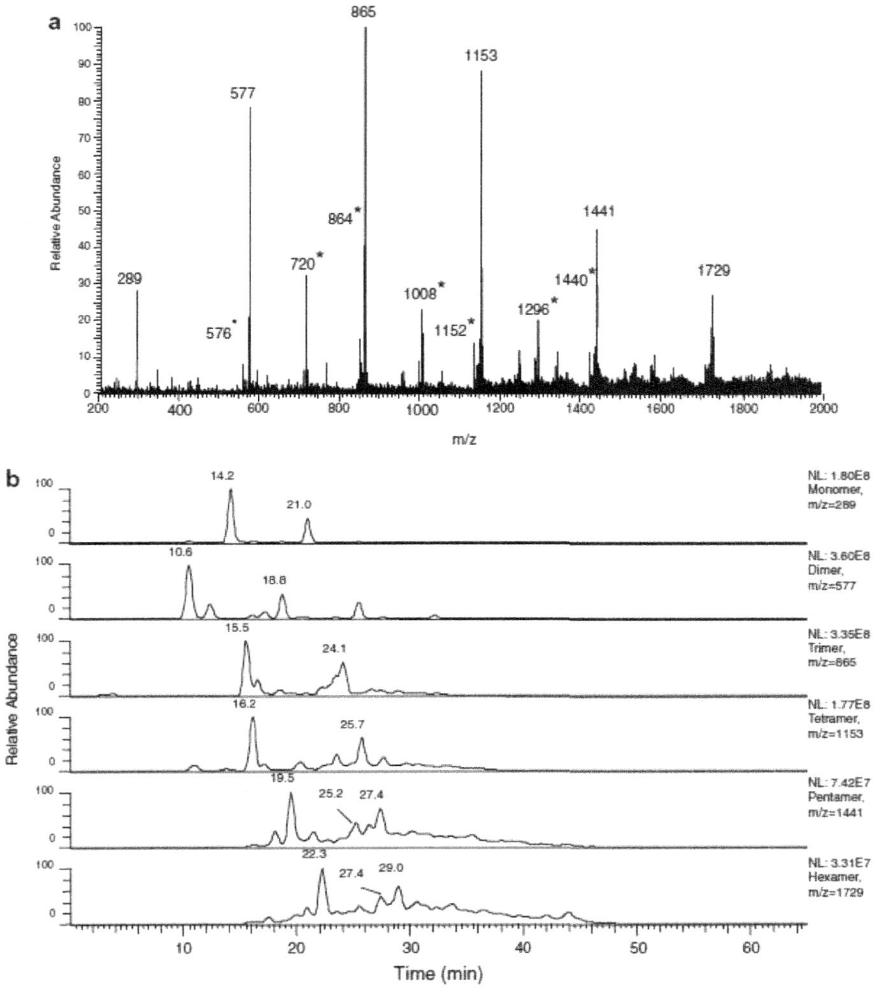

Fig. 2.19 ESI/mass profiles of ANE procyanidins: **a** total ion spectrum and **b** extracted ion chromatography of monomeric to hexameric procyanidin from m/z 289 to 1729 (Huang 2010). Reproduced with permission from Elsevier

Both silver and gold nanoparticles, having surface plasmon resonance in the region of 400–500 nm and 500–550 nm, respectively, could be extracted using areca nut extracts. Both the nanoparticles were able to inhibit the growth of *Candida albicans* with zone of inhibition of 9 mm for silver and 6 mm for gold nanoparticles, suggesting use for pharmaceutical, cosmetic, and other applications (Pradeep 2021). Gold nanoparticles with average size of 15 nm were able to be synthesized using areca nut extracts under room temperature. In this study, areca nut seed extracts (100–400 mg/L) were added to aliquots of 0.2 mM AU (III) solution, which resulted in the

Table 2.17 Inhibitory activity of areca nut extracts on *Staphylococcus aureus*, *Escherichia coli*, and *Salmonella enterica* and *Enterobacter aerogenes* in various solvents by disc diffusion method (Jam 2021)

Solvent	Concentration (mg/mL)	Inhibitory activity of Areca Nut fruit extracts (mm)			
		Enterobacter aerogenes	*Salmonella enterica*	*Escherichia coli*	*Staphylococcus aureus*
H_2O	100	–	15 ± 0.34	15 ± 0.45	–
C_2H_5OH	100	10 ± 0.23	15 ± 0.23	17 ± 1 + 01	16 ± 0.34
CH_3OH	100	12 ± 0.65	16 ± 1.34	19 ± 087	17 ± 0.4
Gentamicin	0.006	20 ± 0.87	30 ± 1.01	35 ± 0.65	25 ± 0.54
Penicillin	0.006	15 ± 0.87	18 ± 0.98	25 ± 0.56	20 ± 0.96

Reproduced from John Wiley and Sons with permission through open access publishing

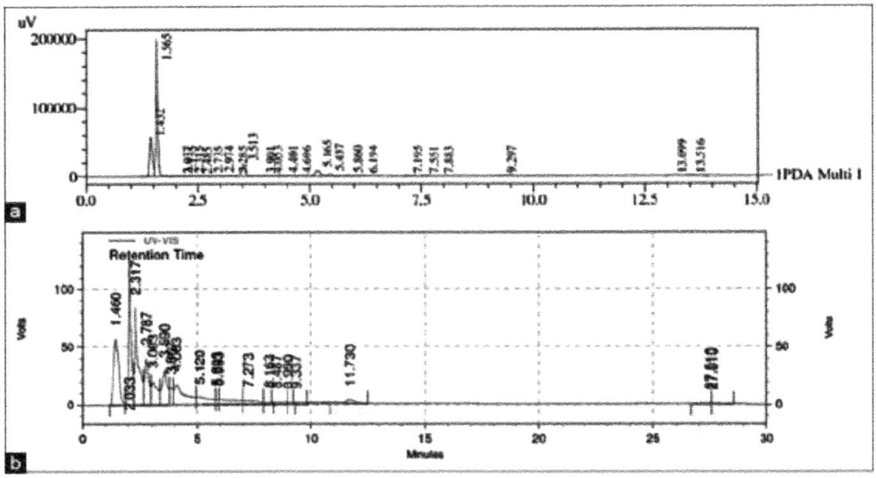

Fig. 2.20 High-performance liquid chromatogram of the areca nut extracts, **a** catechin, **b** quercetin (Sari 2020). Reproduced with permission through open access publishing

formation of nanoparticles. Strong absorption peak was observed at 545 nm in the UV spectrometer curves, confirming the presence of nanoparticles. These nanoparticles had average size of 15 nm and were used as a catalyst for the reduction of 4-nitrophenol to 4-aminophenol in the presence of NaBH (Dash 2019).

Silver nanoparticles were formed when aqueous areca nut extracts were added to silver nitrate and treated between 30 and 95 °C for 10 min. Particles with average size of 80 nm were obtained and showed significant antitumor activity when injected into mice infected with Dalton's Ascites Lymphoma (Sukritha 2011). Green synthesis of silver nanoparticles having diameter of 12 nm was reported using aqueous areca nut extracts at 60 °C for 70 min. These particles were able to photodegrade methylene

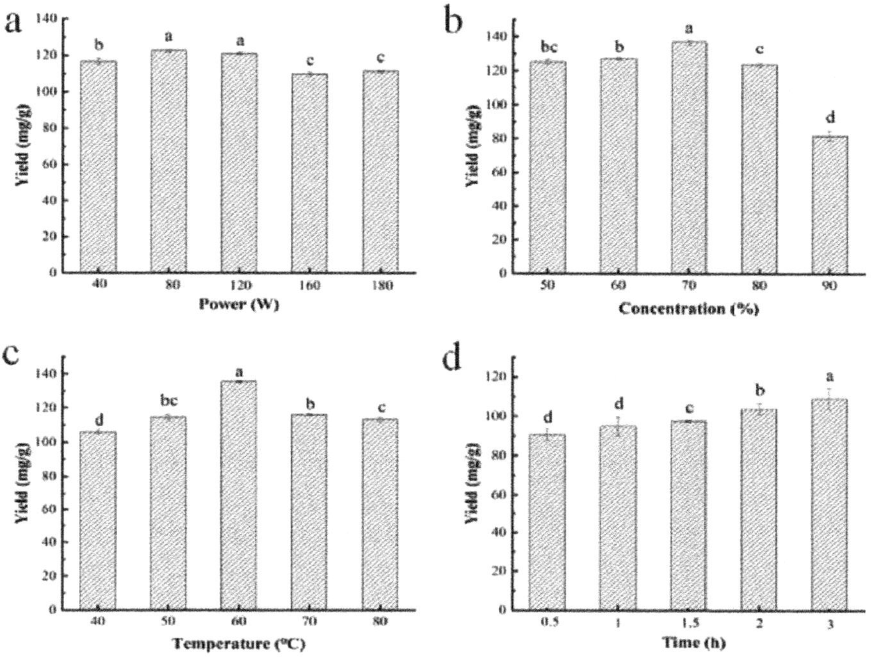

Fig. 2.21 Effect of ultrasonic power (**a**), ethanol concentration (**b**), temperature (**c**), and time (**d**) on the extraction yield of ACP. Values are expressed as means ±SD ($n = 3$). Different lowercase letters (**a**–**d**) in the same figure indicate a statistically significant difference ($p < 0.05$) (Sun 2023). Reproduced with permission through open access publishing

blue in water and also showed high microbial activity against both gram-positive and gram-negative bacteria (Vinay 2019).

In a different approach, hydrotalcite-silver (HT-Ag) nanosheets (Fig. 2.26) were synthesized using $AgNO_3$ salt solution as precursor and areca nut extract as the reducing agent through co-precipitation and hydrothermal treatments. Nanoparticles obtained were in the range of 15–40 nm (Fig. 2.27) with high yield of 61–86%. Further, the nanoparticles showed excellent antibacterial properties against *E. coli* and *S. aureus,* and even *P. stutzeri B27*. Areca nut extract was used to mediate the production of TiO_2 nanoparticles and was found to have antioxidant, antimicrobial, and antituberculosis properties. The nanoparticles obtained were considered suitable for synthesis of biscoumarin derivatives and exhibited antioxidant properties against butylated hydroxytoluene (BHT) (Nguyen 2024; Kumar 2022).

2.4.1.2 Gold Nanoparticles

Aqueous extracts of areca nut were able to act as a reducing agent and assist in the synthesis of gold nanoparticles (colloids). Both microwave and conventional

Table 2.18 Identified compounds in analyzed samples obtained using HS-SPME and SHDE of areca nuts (Machova 2021)

Group of compounds volatile compounds	Retention index	CAS number	Relative % of Peak Area			
			HS-SPME extraction	SHDE extract	Distillation residue	Hydrolate
ALKANES						
Hexadecane	1597	544-76-3	0.16	–	–	–
Nonadecane	1897	629-92-5	0.31	–	–	–
∑ALKANES			0.47 (2)	–	–	–
ALCOHOLS						
1-Heptanol	972	111-70-6	–	0.07	–	–
1-Octen-3-ol	981	3391-86-4	–	0.05	–	–
3-Octanol	998	589-98-0	–	0.01	–	–
Benzenemethanol	1034	100-51-6	0.56	–	–	–
1-Octanol	1073	111-87-5	0.18	0.13	–	–
Benzeneethanol	1111	60-12-8	0.31	–	–	–
1-Methyl-4-(1-methylethyl)-cis-2-cyclohexen-1-ol	1125	29803-82-5	–	0.01	–	–
1-Hexadecanol	1878	36653-82-4	0.34	–	–	–
∑ ALCOHOLS			1.39 (4)	0.27 (5)	–	–
ALDEHYDES						
trans-2-Heptenal	957	18829-55-5	–	0.01	–	–
Benzaldehyde	960	100-52-7	1.02	0.05	–	7.23
Octanal	1003	124-13-0	–	0.16	–	–
Phenylacetaldehyde	1042	122-78-1	0.56	0.01	–	–
Nonanal	1105	124-19-6	2.45	0.38	–	–

(continued)

Table 2.18 (continued)

Group of compounds volatile compounds	Retention index	CAS number	Relative % of Peak Area			
			HS-SPME extraction	SHDE extract	Distillation residue	Hydrolate
trans-2-Nonenal	1160	18829-56-6	0.37	0.09	–	–
Decanal	1205	112-31-2	0.40	0.07	–	2.16
p-Isopropylbenzaldehyde	1241	122-03-2	0.23	–	–	–
Anisaldehyde isomer	1254	–	0.17	–	–	–
trans-2-Decenal	1263	3913-81-3	0.24	0.15	–	–
Undecanal	1306	112-44-7	0.19	–	–	–
trans-2, trans-4-Decadienal	1319	25152-84-5	0.17	0.14	–	–
2-Undecenal	1364	2463-77-6	–	0.04	–	–
2-Butyl-2-octenal	1371	13,019-16-4	–	0.13	–	–
Tridecanal	1509	10486-19-8	0.22	–	–	–
Tetradecanal (Myristaldehyde)	1610	124-25-4	0.20	–	–	–
1-Pentadecanal	1712	2765-11-9	0.22	–	–	–
∑ALDEHYDES			6.44 (13)	1.23 (11)	–	9.39 (2)
ESTERS						
Amyl acetate	915	628-63-7	–	0.01	–	–
Dodecanoic acid, methyl ester	1521	111-82-0	0.75	–	–	–
Dodecanoic acid, ethyl ester	1591	106-33-2	6.80	–	–	–
Tetradecanoic acid, methyl ester	1721	124-10-7	0.96	–	–	–
Tetradecanoic acid, ethyl ester	1789	124-06-1	6.04	–	–	–
Tetradecanoic acid, 1-methylethyl ester	1820	110-27-0	0.51	–	–	–

(continued)

Table 2.18 (continued)

Group of compounds volatile compounds	Retention index	CAS number	Relative % of Peak Area			
			HS-SPME extraction	SHDE extract	Distillation residue	Hydrolate
Amyl laurate	1878	5350-03-8	–	–	0.82	–
Hexadecanoic acid, methyl ester	1927	112-39-0	0.22	–	–	–
Hexadecanoic acid, ethyl ester	1989	628-97-7	0.61	–	–	–
Hexadecanoic acid, 1-methylethyl ester	2019	142-91-6	4.88	–	–	–
∑ ESTERS			20.77 (8)	0.03 (2)	0.82 (1)	–
FATTY ACIDS						
Hexanoic acid	982	142-62-1	–	–	–	7.65
Heptanoic acid	1093	111-14-8	–	0.03	–	–
Octanoic acid	1172	124-07-2	0.61	–	–	20.0
Nonanoic acid	1239	112-05-0	0.19	2.48	–	–
Decanoic acid	1386	334-48-5	1.47	0.08	0.13	–
Undecanoic acid	1436	112-37-8	–	0.09	–	–
Dodecanoic acid	1560	143-07-7	60.40	45.23	17.88	18.1
Tetradecanoic acid	1752	544-63-8	–	0.51	63.24	2.78
Pentadecanoic acid	1840	1002-84-2	–	28.56	–	–
Hexadecanoic acid	1944	57-10-3	0.36	0.16	12.43	–
Heptadecanoic acid	2036	506-12-7	–	4.01	–	–
cis-9, cis-12-Octadecadienoic acid	2129	60-33-3	–	–	0.88	–
trans-9-Octadecenoic acid	2135	112-80-1	0.48	4.07	1.99	–
∑ FATTY ACIDS			63.51 (6)	85.22 (10)	96.55 (6)	48.53 (4)

(continued)

Table 2.18 (continued)

Group of compounds volatile compounds	Retention index	CAS number	Relative % of Peak Area			
			HS-SPME extraction	SHDE extract	Distillation residue	Hydrolate
KETONES						
Oct-3-en-2-one	1038	1669-44-9	–	0.07	–	–
Acetophenone	1065	98-86-2	–	0.01	–	–
2-Undecanone	1292	112-12-9	–	0.20	–	–
Nerylacetone	1445	3879-26-3	0.14	–	–	–
2-Tridecanone	1495	593-08-8	1.52	0.07	–	–
3-ethenyl-3-methyl-6-(1-methylethyl)-2-(1-methylethylidene)-cyclohexanone	1528	21698-46-4	–	0.14	–	–
γ-Dodecalactone	1675	2305-05-7	0.50	–	–	–
2-Pentadecanone	1694	2345-28-0	2.43	–	–	–
∑ KETONES			4.59 (4)	0.49 (5)	–	–
TERPENES and TERPENOIDS						
α-Thujene	925	2867-05-2	–	0.02	–	–
α-Pinene	932	80-56-8	–	0.06	–	–
Camphene	948	79-92-5	–	0.01	–	–
β-Pinene	976	127-91-3	–	0.06	–	–
α-Terpinen	1016	99-86-5	–	0.06	–	–
p-Cymene	1023	99-87-6	–	0.16	–	–
Limonene	1029	138-86-3	–	0.17	–	–
Eucalyptol	1031	470-82-6	–	0.11	–	4.11
β-Ocimene	1046	13877-91-3	–	0.01	–	–

(continued)

Table 2.18 (continued)

Group of compounds volatile compounds	Retention index	CAS number	Relative % of Peak Area			
			HS-SPME extraction	SHDE extract	Distillation residue	Hydrolate
cis-Linalool oxide	1071	5989-33-3	–	0.02	–	4.10
Terpinolene	1085	586-62-9	–	0.02	–	–
trans-Linalool oxide	1087	34995-77-2	–	–	–	4.19
Fenchone	1088	1195-79-5	–	0.11	–	–
Linalool	1101	78-70-6	–	0.41	–	2.73
Thujone	1117	546-80-5	–	0.03	–	–
Camphor	1147	76-22-2	–	0.43	–	5.38
Isomenthone	1156	491-07-6	–	0.28	–	–
D-isomenthone	1165	1196-31-2	–	0.18	–	–
Menthol	1171	89-78-1	–	0.02	–	–
L-Borneol	1173	464-45-9	–	0.15	–	–
Terpinen-4-ol	1182	562-74-3	–	0.71	–	5.60
α-Terpineol	1196	98-55-5	–	0.40	–	3.01
D-Carvone	1246	2244-16-8	–	0.46	–	6.06
Geraniol	1253	106-24-1	–	0.02	–	–
Piperitone	1255	89-81-6	–	0.04	–	–
α-Terpinyl acetate	1347	80-26-2	–	0.02	–	–
Geranyl acetone	1448	3796-70-1	–	0.03	–	–
α-Curcumene	1482	644-30-4	–	0.35	–	–
β-Selinene	1491	17066-67-0	–	0.08	–	–

(continued)

Table 2.18 (continued)

Group of compounds volatile compounds	Retention index	CAS number	Relative % of Peak Area			
			HS-SPME extraction	SHDE extract	Distillation residue	Hydrolate
α-Selinene	1498	473-13-2	–	0.05	–	–
β-Bisabolene	1508	495-61-4	–	0.03	–	–
Myristicin	1522	607-91-0	–	0.22	–	–
Elemicin	1547	487-11-6	–	0.02	–	–
∑ TERPENES and TERPENOIDS			–	4.74 (32)	–	35.18 (8)
Other compounds						
2,6,6-trimethyl-2-ethenyltetrahydropyran	969	7392-19-0	–	0.01	–	–
2-pentylfuran	989	3777-69-3	–	0.16	–	–
4-ethenyl-1,5,5-trimethylcyclopentene	1058	1727-69-1	–	0.16	–	–
4-Acetyl-1-methylcyclohexene	1130	6090-09-1	–	0.01	–	–
Arecoline	1225	63-75-2	0.80	–	1.85	–
Anethole	1287	4180-23-8	0.76	0.42	–	–
6-Nonyltetrahydro-2H-pyran-2-one	1950	2721-22-4	–	0.05	–	–
∑ OTHER COMPOUNDS			1.56 (2)	0.83 (7)	1.85 (1)	

(∑ in table is expressed always for each group of compounds; first number is the sum of % rel. and the second number in () is number of compounds)
Reproduced with permission through open access publishing

Table 2.19 Effect of oral areca catechu consumption on wound contraction rate and epithelialization period of burn wound model in rats (Bharat 2014)

Group	Percentage of wound contraction (Mean ± SEM)								Period of epithelialization (days)
	Day 3	Day 5	Day 7	Day 9	Day 11	Day 13	Day 15	Day 17	
Control	5.74 ± 0.34	9.2 ± 0.58	15.35 ± 0.55	27.04 ± 0.49	39.20 ± 0.56	52.20 ± 0.82	65.35 ± 0.99	74.61 ± 1.26	23.67 ± 0.42
Standard	14.25 ± 5.02	23.91 ± 5.70	29.76 ± 5.82	37.56 ± 5.96	61.62 ± 5.55	81.10 ± 5.28	91.11 ± 3.10	100 ± 0.00	15.67 ± 0.67
100 mg/kg extract	9.16 ± 1.50	17.88 ± 2.46	26.26 ± 2.92	35.36 ± 2.96	52.34 ± 2.96	72.34 ± 2.76	93.14 ± 2.22	100 ± 0.00	16.33 ± 0.42
300 mg/g extract	15.87 ± 3.33	28.99 ± 2.08	37.76 ± 1.27	47.08 ± 1.20	62.19 ± 1.96	77.08 ± 1.20	92.85 ± 1.20	98.24 ± 1.12	17.67 ± 0.41

Reproduced with permission through open access publication

Table 2.20 Effect of oral areca catechu consumption on wound contraction rate and epithelialization period of dexamethasone delayed burn wound model in rats (Bharat 2014)

Group	Percentage of wound contraction (Mean ± SEM)									Period of epithelialization (days)
	Day 3	Day 5	Day 7	Day 9	Day 11	Day 13	Day 15	Day 17	Day 19	Day 21	
Dexa	3.20 ± 0.17	6.41 ± 0.40	12.61 ± 0.59	19.82 ± 0.63	24.74 ± 0.67	37.82 ± 0.86	50.47 ± 1.29	66.39 ± 1.13	79.24 ± 1.30	84.82 ± 1.02	28.33 ± 0.42
Dexa + 100 mg extract	6.87 ± 0.58	14.07 ± 0.87	22.76 ± 1.16	27.71 ± 1.13	34.15 ± 0.99	51.39 ± 1.42	67.66 ± 2.38	82.93 ± 2.26	94.27 ± 1.91	100 ± 0.00	20.33 ± 0.42
Dexa + 300 mg extract	14.00 ± 1.73	23.36 ± 1.65	31.76 ± 1.10	35.88 ± 0.63	42.59 ± 1.22	57.39 ± 0.97	69.94 ± 1.90	83.70 ± 2.57	93.17 ± 1.70	99.29 ± 0.71	21.33 ± 0.33

Reproduced with permission through open access publication

Table 2.21 Surface textural properties of areca nut carbons (ANCs) obtained from nitrogen sorption analysis (Joshi 2021)

Sample	SSA ($m^2\ g^{-1}$)	S_{mic} ($m^2\ g^{-1}$)	V_p ($cm^3\ g^{-1}$)	V_{mic} ($cm^3\ g^{-1}$)	$V_{meso}/V_p \times$ 100 (%)	W_p (nm)	D_p (nm)
ANC0.25_400_3h	1399.9	1008.8	1.464	0.925	36.81	0.704	3.67
ANC0.5_400_3h	1705.0	1124.7	2.088	1.257	39.79	0.705	3.68
ANC1_400_3h	2132.1	1320.8	3.426	1.920	43.95	0.705	3.29
ANC1.5_400_3h	1945.2	1219.3	3.152	1.791	43.17	0.705	3.30
ANC1_400_2h	1970.4	1278.1	2.643	1.563	40.86	0.704	3.67
ANC1_400_4h	1781.3	1187.5	2.333	1.393	40.29	0.705	3.68
ANC1_300_3h	1156.1	903.4	1.112	0.709	36.24	0.286	3.66
ANC1_500_3h	1934.8	1217.4	2.878	1.635	43.18	0.705	3.49

SSA, specific surface area; S_{mic}, micropore surface area; V_p, total pore volume; V_{mic}, micropore volume; V_{meso}, mesopore volume; W_p, average half pore width; and D_p, average mesopore diameter
Reproduced with permission through open access publishing

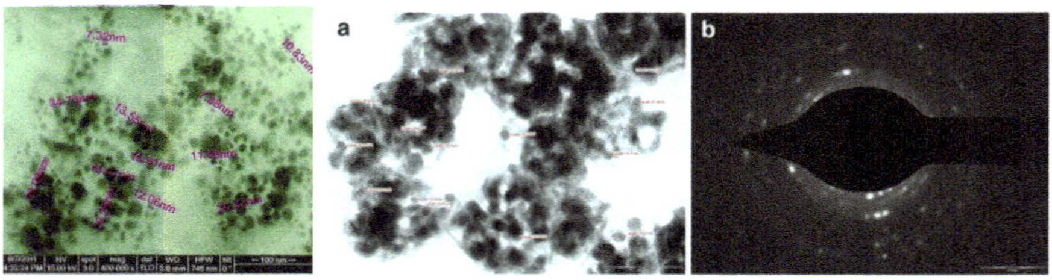

Fig. 2.22 TEM and FESEM images of silver nanoparticles synthesized using areca nut extracts (Bhat 2013). Reproduced with permission through open access publishing

synthesis approaches were used with a frequency of 2450 MHz for 1 min and temperatures at 300°K and 373°K, respectively. Size of nanoparticles (13.7–22.2 nm) (Fig. 2.28) obtained varied depending on the process and temperature used during the synthesis, and average crystallite size obtained was 14 nm. Flavonoids and proteins present in the extract were considered to be responsible for the reducing potential of the areca nut solution. The nanoparticles were able to remove hazardous pollutants, including dyes, had excellent antioxidant activity and antimicrobial and anticancer properties, suggesting suitability for environmental and biomedical applications (Rajan 2015).

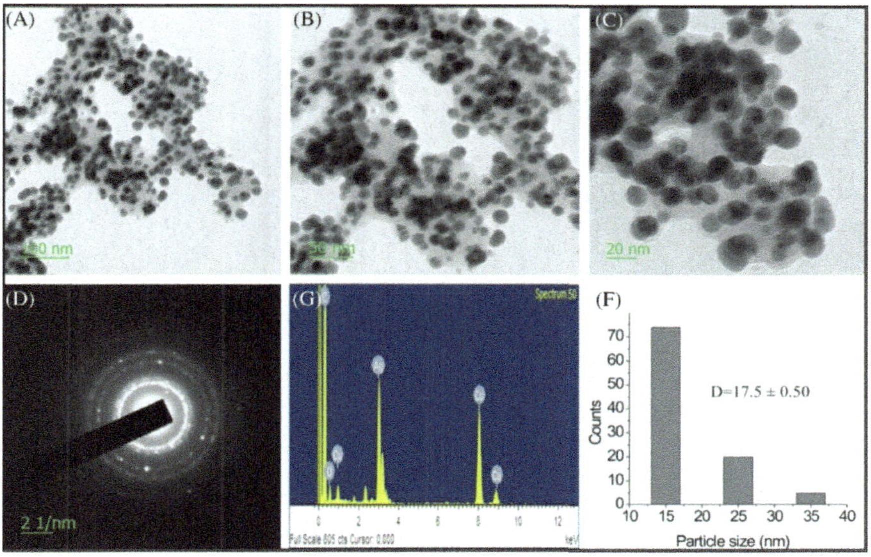

Fig. 2.23 **a–c** HRTEM images of AgNPs. **d** SAED images of AgNPs. **e** EDX spectra of AgNPs. **f** Particle size distribution plot (Dash 2022). Reproduced with permission from Springer Nature through open access publishing

2.4.1.3 Zinc Nanoparticles

To synthesize nanoparticles, areca nut extract was added to zinc nitrate solution and reacted for one hour to collect the supernatant. Later, calcination was done at 400 °C for 120 min, resulting in the formation of ZnO nanoparticles. During the reaction, color of the solution changed from dark wine to white after calcination confirming the formation of the nanoparticles (Fig. 2.29). Absorption spectroscopy also showed the characteristic peak at 346 nm (Raghavendra 2022). Size of the nanoparticles was between 5 and 30 nm, and zeta potential was +7.1 mV. X-ray diffraction measurements showed peaks at 31.84°, 34.52°, 36.33°, 47.63°, 56.64°, 62.91°, 66.54°, and 69.12° representing the (100), (002), (101), (102), (110), (103), (200), (112) and (201) planes which confirm the wurtzite structure of ZnO nanoparticles. The nanoparticles were able to degrade plastic and also dyes, making them useful for environmental remediation and medical applications.

2.4.1.4 Synthesis of Palladium Nanoparticles

In addition to silver and gold, areca nut extracts have been used for synthesis (Fig. 2.30) of other metallic nanoparticles. In one such example, the phytochemicals (amines, alcohols, and phenols) in areca nut were used for the conversion of Pd (II) into Pd (O) and obtain palladium nanoparticles without the need for any external

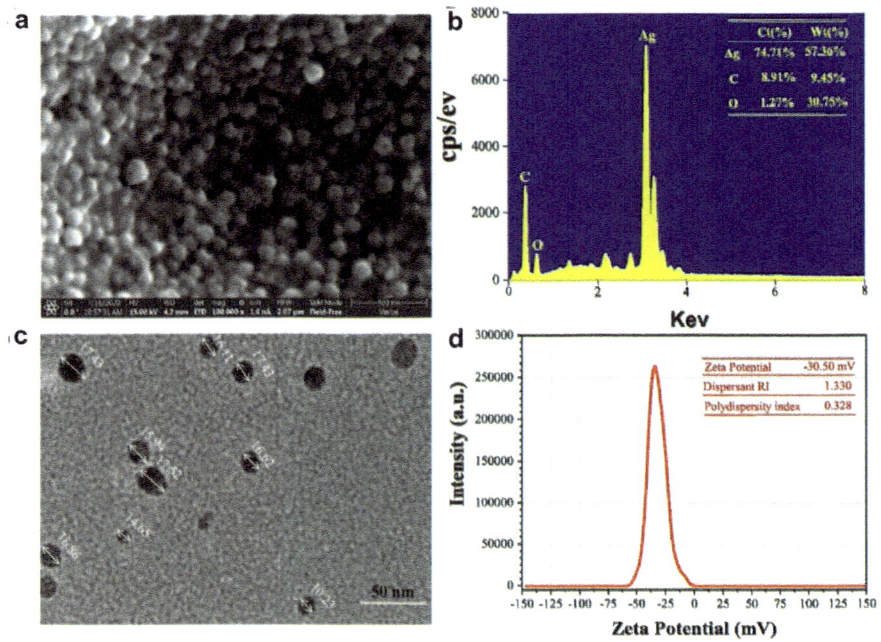

Fig. 2.24 SEM image (100,000×) (**a**), EDX spectrum (**b**), TEM image (**c**), and Zeta potential value (**c**) of the purified AgNPs synthesized under the optimal synthesis parameters (pH value at 12.0 of the extract; 1.0 mM of $AgNO_3$ concentration; 90 min of reaction time) (Hu 2022). Reproduced with permission through open access publishing

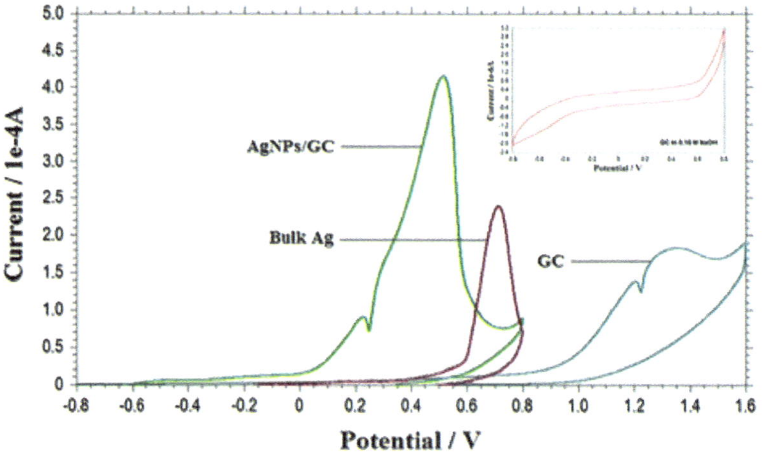

Fig. 2.25 Cyclic voltammogram of GC, bulk Ag and AgNPs/GC toward the electrocatalytic oxidation of 5 mM glucose in 0.10 M aqueous NaOH; *inset* cyclic voltammogram of GC in 0.10 M NaOH (Edison 2017). Reproduced from Springer Nature with permission through open access publishing

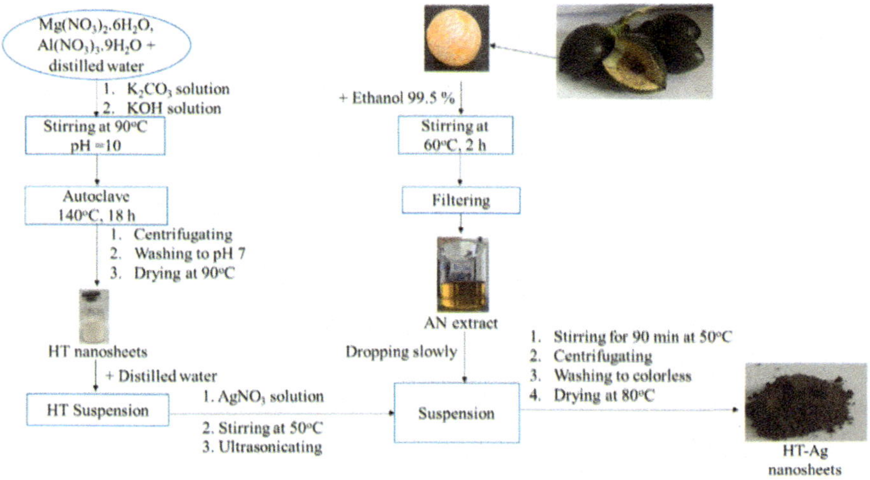

Fig. 2.26 Schematic representation of the process used to develop the HT-Ag nanosheets using areca nut extracts (Nguyen 2024). Reproduced with permission through open access publishing

ligands. Nanoparticles obtained were spherical in shape with average size less than 100 nm (Hegde 2021). Surface area of the particles was about 8.78 m^2/g, and mean pore diameter was 16.7 nm. The nanoparticles could be used as catalyst for synthesis of several chemicals.

In another study, chitosan nanoparticles having average size of 254 nm, zeta potential of 12.7, and polydispersity index of 0.436 were obtained when extracts obtained from areca nut using 50% ethanol were used as the reducing agents (Rahman 2023). In a recent study, phytochemicals in areca nuts were used to produce pyrimidine derivatives for pharmaceutical applications. For this, areca nut extracts were used to synthesize δ-MnO_2 nanoparticles with antimicrobial activity against gram-negative bacteria (Limaye 2024).

2.4.2 *Nanocellulose from Areca Nuts*

Instead of preparing nanoparticles, areca nuts were used to extract nanocellulose, and the extracted nanocellulose was studied as the potential carrier for quercetin (Q_T). In this approach, areca nuts were defatted and later treated with 6% NaOH solution at 70 °C for 2 h. Subsequently, the samples were treated to remove lignin and later subject to acid hydrolysis to separate minerals and amorphous cellulose. Further treatment with 1, 10, or 15% sulfuric acid resulted in nanocellulose. The obtained nanocellulose was combined with zinc (Zn) ions, and the ability of the nanocellulose to load and release quercetin under different conditions was studied. Combining QT with Zn arecanut nanocellulose increased the antibacterial activity

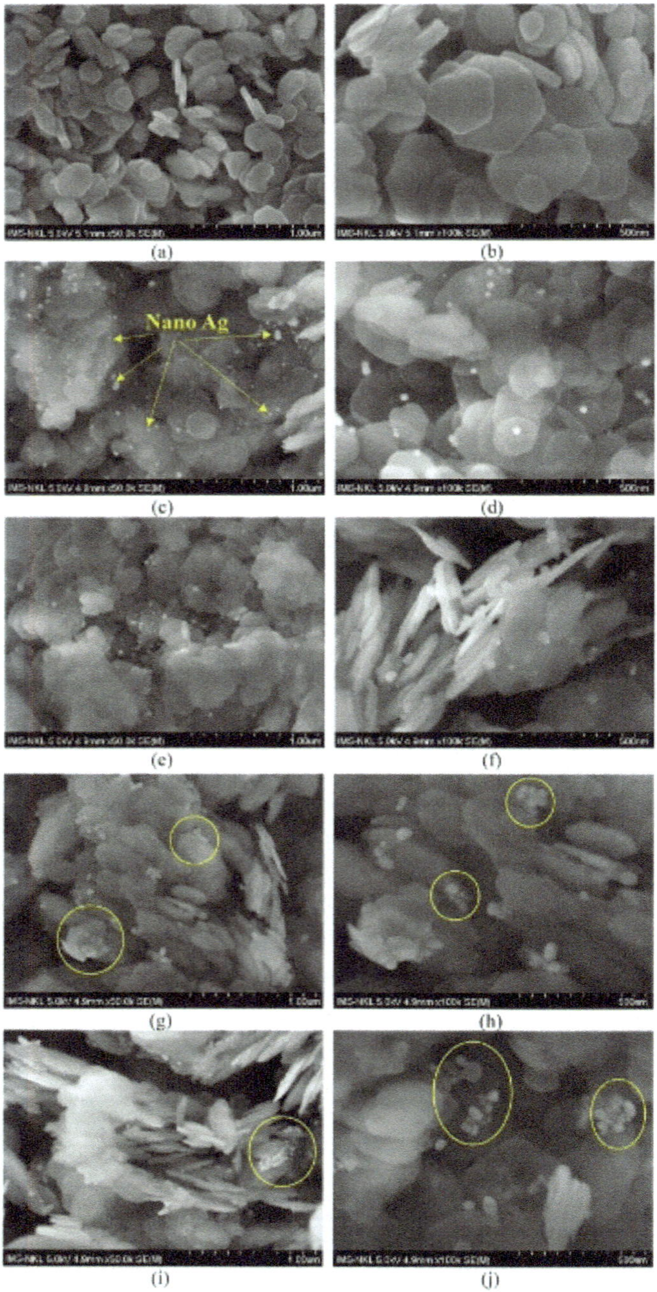

Fig. 2.27 SEM images of HT (**a**, **b**), HT-Ag5 (**c**, **d**), HT-Ag10 (**e**, **f**), HT-Ag15 (**g**, **h**), HT-Ag20 (**i**, **j**) at a magnification of 50 and 100 k. Yellow circles present the clusters of agglomerated Ag NPs on HT surface (Nguyen 2024). Reproduced with permission through open access publishing

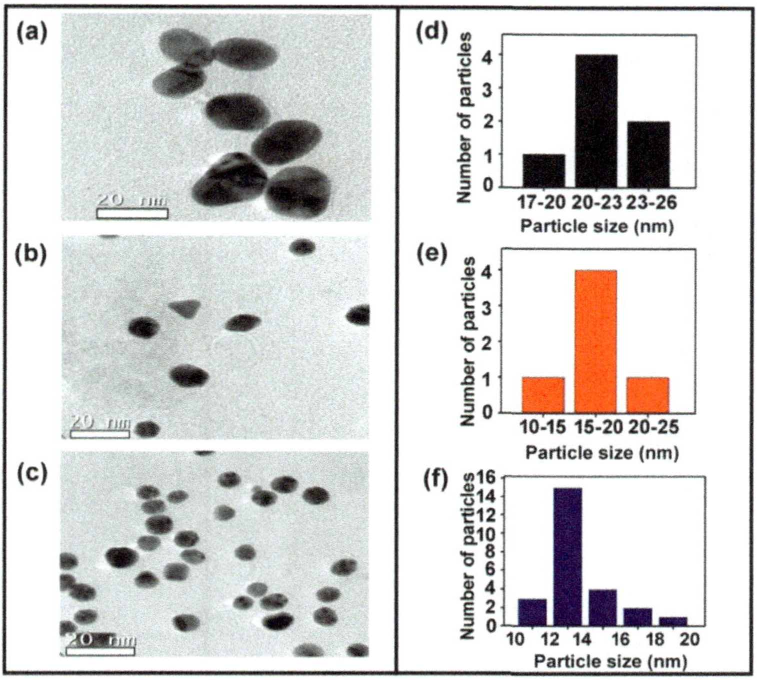

Fig. 2.28 TEM images of colloids (**a**) A, (**b**) B, (**c**) C, and their particle size distribution histograms (**d**) A, (**e**) B, and (**f**) C, respectively (Rajan 2015). Reproduced with permission from Elsevier

and shelf life. As seen from Fig. 2.31, a sustained release rate was possible, and the drug-Zn-nanocellulose complex was considerably stable since the zinc ions shielded the drug from the external environment (Li 2024).

2.4.3 Environmental Remediation Using Areca Wastes

Considerable portion of arecanuts remains as waste after processing. This areca waste has been considered for various applications, including environmental remediation. In a simple approach, areca nut obtained as waste was directly used as a sorbent for removal of copper and cadmium in water, and the sorption properties were studied under different conditions. High sorption potential of 97.75% for Cd^{2+} and 96.46% for Cu^{2+} were obtained even at a low sorbent dosage of 1 g/100 mL (Fig. 2.32). pH of the solution was found to be most critical factor with sorption occurring through surface adsorption and ion-exchange (Zheng 2008). Instead of using the areca nuts or extracts, the potential of areca nut waste obtained after chewing to act as an absorbent for Pb^{2+} in water was studied. The waste was treated with Fenton reagent to activate the surface before using it for absorption. Absorption studies were done

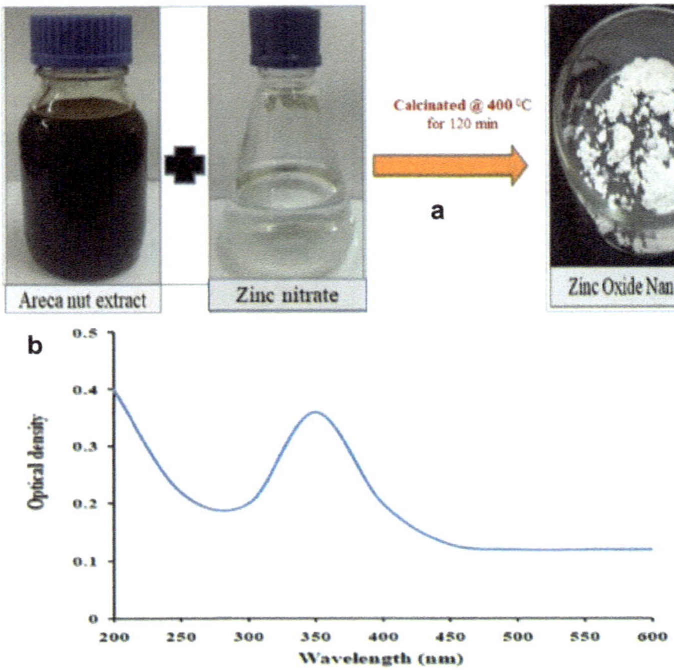

Fig. 2.29 Areca nut water extract with zinc nitrate, thus, forming the color change from dark wine color to white precipitate **B** UV–visible spectra obtained for biosynthesized ZnO NPs were observed at 346 nm (Raghavendra, 2022). Reproduced with permission from Springer nature

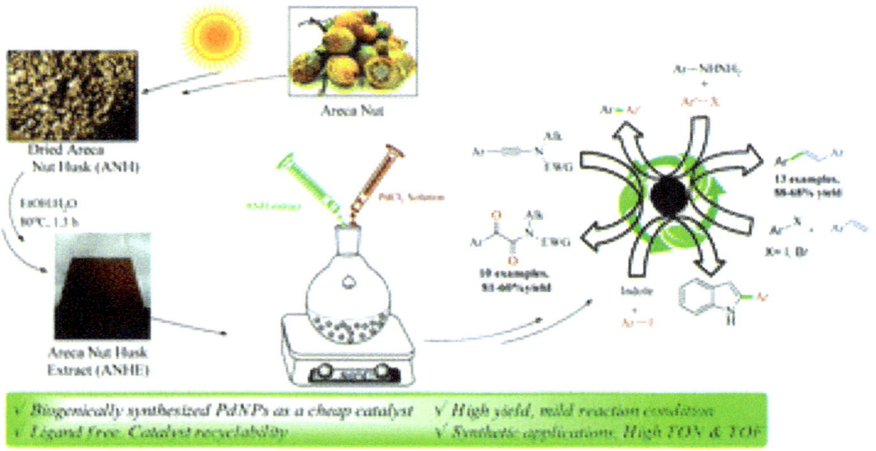

Fig. 2.30 Graphical representation of the process used to synthesize the palladium nanoparticles using areca nut extracts (Hegde 2021). Reproduced with permission from Royal Society of Chemistry

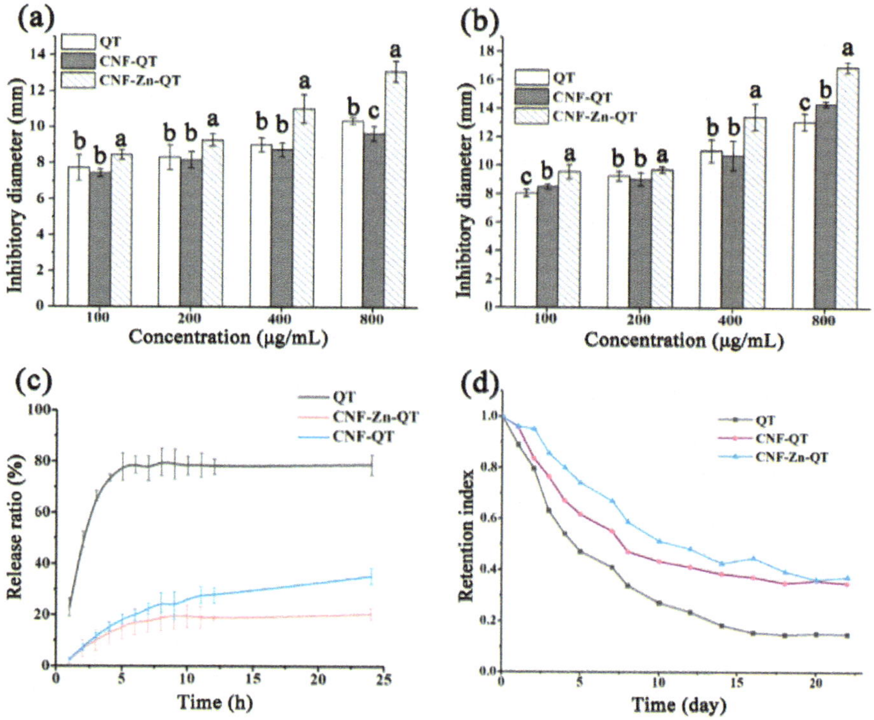

Fig. 2.31 Antibacterial activity against *Escherichia coli* (**a**) and *Staphylococcus aureus* (**b**); in vitro release rate of the samples during 24 h (**c**) at different concentrations; the retention index of QT in different samples over 22 days (**d**) (Li 2024). Reproduced with permission through open access publishing

using initial Pb^{2+} concentrations of 4–12 mg/L and adsorbent dosage of 1 g. As seen from Fig. 2.33, the removal efficiency was dependent on the initial concentration and sorption time. Sorption was found to follow both Langmuir and Freundlich models, and the process closely followed pseudo-second order kinetics. A negative Gibbs free energy and positive enthalpy indicated spontaneous and endothermic absorption. The entropy during the reaction was also positive suggesting affinity between the areca waste and Pb^{2+} (Li 2010).

In a similar approach, areca nut waste was pyrolyzed into activated carbon and later functionalized with succinic anhydride under microwave irradiation for potential use as a sorbent for removal of phenol and mercury (Hg(II)). Specific surface area of the activated and functionalized carbon was 308 m^2/g, which provided maximum sorption capacity of 11.23 and 5.37 mg/g for Hg(II) and phenol, respectively. The modified carbon was considered to be suitable for the simultaneous removal of Hg(II) and phenol in wastewater (Lalhmunsiama 2017).

In another attempt to use areca nuts for environmental remediation, novel porous nanospheres were synthesized through a one-step pyrolysis. The sorbent obtained

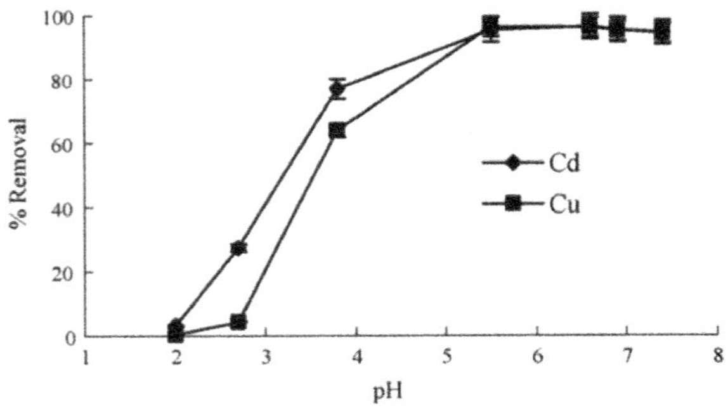

Fig. 2.32 The effect of pH on metal removal through areca nut waste adsorption (amount of AW = 10 g/L, agitation time is 120 min, metal concentration 8 mg/L) (Zheng 2008). Reproduced with permission from Elsevier

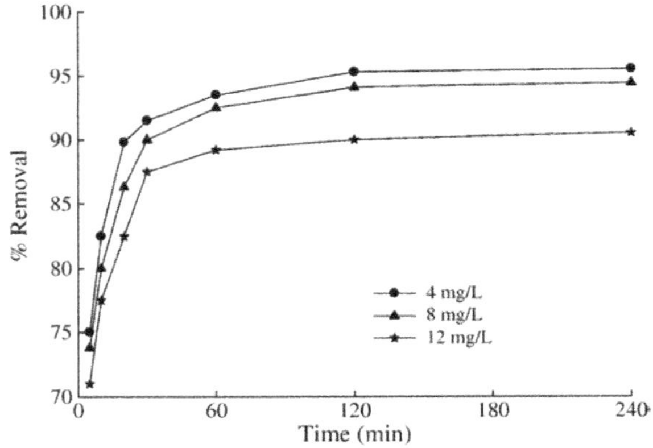

Fig. 2.33 Effects of initial concentration and time on sorption of Pb^{2+} (Li 2020). Reproduced with permission from Elsevier

was used for the removal of methylene blue and anionic methyl orange and their binary mixtures. Areca nut carbon had high specific surface area of 693 m^2/g and highly microporous nature when prepared at 1000 °C. A removal efficiency higher than 95% for the dyes and their blends was obtained even at a low sorbent concentration of 0.15 g/L and sorption time of 70 min. Highest sorption was 97 mg/g for methylene blue, 71 mg/g for orange, and 50 mg/g for their blends. Sorption was mostly through electrostatic and π–π interactions between the dyes and areca nut nanospheres (Pathania 2023). Activated carbon was prepared from areca nuts under various processing conditions and treated with H_3PO_4 as the activating agent for potential use as a sorbent for methylene blue in wastewater. Yield of the carbon was between 40 and 45%, depending on the carbonization conditions. Some of the properties of the carbon obtained are given in Table 2.21. The extent of methylene blue sorption was also dependent on the sorption conditions and the properties of the carbon. A high sorption of 333 mg of dye per gram of carbon was possible due to the properties of the activated carbon (Joshi 2021). In a similar study, porous nanospheres were prepared from areca nuts using a one-step pyrolysis at 1000 °C. The carbon obtained had specific surface area of 694 m^2/g with highly microporous structure. When used as a sorbent for dyes, the carbon was able to provide higher than 95% efficiency with sorption capacity of 97 mg/g for methylene blue and 71 mg/g for methyl orange due to π–π interactions between the dye molecules and carbon, and hence considered suitable for large-scale effluent treatment in textile dyeing industries (Pathania 2023).

In a different approach, arecanut extract was combined with chitosan and PVA and made into a film (Fig. 2.34) for removal of chromium (VI) from water. Depending on the amount of extract in the films, the BET surface area of the films ranged from 37 to 51 m^2/g compared to about 30 m^2/g for pure chitosan and PVA films. The films containing 0.4 g of areca extract were able to remove 88% of chromium (173 mg/L) when used at pH 9.0, 40 °C for 45 min. Sorption was mostly through pseudo-second order reactions (Jose 2024). In a related study, areca waste was treated with H_3PO_4 and later carbonized and used as a biosorbent for Cr(VI). Chemical modification provided enhanced sorption potential with highest sorption capacity of 188 mg/g at pH 2. The sorbent was able to eliminate even trace levels of the metal present individually or as a mixture with other metals or metal ions. About 94% of Cr (VI) could be easily desorbed after treating with 2 M NaOH, and hence the sorbent was considered suitable for industrial level waste water treatment (Basnet 2025). In another study, areca nut treated with concentrated sulfuric acid were shown to have good potential for removal of Cr(VI). Extent of sorption was dependent on pH and temperature. High sorption of 2 mmol per g was obtained through a spontaneous sorption process and endothermic reaction. Treated nuts also had good desorption potential and could be reused multiple times (Basnet 2022).

The possibility of converting areca nuts into carbon and using the obtained carbon as a sorbent for carbon dioxide was studied by Supriya, et al. Areca nuts were directly pyrolyzed into carbon with high porosity without using any activating agent. Depending on the carbonization temperature, the carbon had specific surface area between 433 and 1000 m^2/g and high carbon content between 96 and 99%. The carbon obtained was able to sorb 14.1 mmol/g of CO_2 at 0 °C and 30 bar, which was considered to be high compared to carbons obtained from activated carbon and carbon nanotubes (Supriya 2022). In another attempt, the potential of using areca nut for carbon capture and preparation of supercapacitors was evaluated. Initially, areca nuts were carbonized by heating at 500 °C for 2 h under nitrogen atmosphere. A carbon yield of 78% was obtained with composition of 63wt% C, 1.83% N, 2.64% H, and 32.7% O. This carbon was activated by treating with KOH solution and heating between 600 and 800 °C for 2 h under nitrogen atmosphere, with yields ranging from 68 to 90%. Activated carbon formed was doped with nitrogen, resulting in carbon containing 65.72 wt.% C, 0.83 wt.% N, 3.57 wt.% H, and 29.88 wt.% O. Some of the properties of the areca nut carbon obtained after various treatments are shown in Table 2.22. Considerably high CO_2 capturing ability between 4.3 and 4.8 mmol/g was obtained depending on the temperature during synthesis. Similarly, specific capacitance was high between 315 and 362 F/g, which increased to 419 F/g after doping with nitrogen, with good cyclic stability of up to 1000 and capacitance retention higher than 91% (Chou 2020).

In another approach, areca nuts were pyrolyzed at 550 °C under nitrogen atmosphere to form carbon nanospheres. The spheres obtained have a diameter between 30

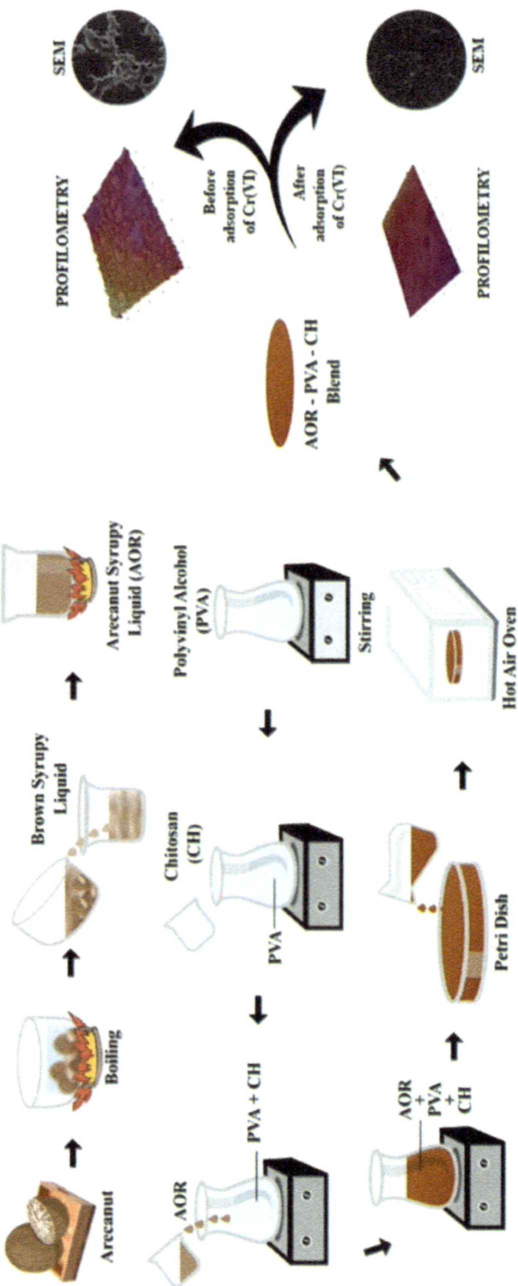

Fig. 2.34 Schematic of the process used to develop and characterize the arecanut extract containing PVA-chitosan developed as sorbent for removal of chromium (Jose 2024). Reproduced with permission through Elsevier

Table 2.22 Properties of various carbons obtained from areca nut (Chou 2020)

Physical parameters					Elemental analysis					
Sample	S_{BET}[a] (m^2/g)	V_{micro}[b] (cm^3/g)	V_{total}[c] (cm^3/g)	V_{micro}/V_{total} (%)	V_{meso}/V_{total} (%)	Pore size (nm)	Found C (%)	Found N (%)	Found H (%)	Found O (%)
AN	–	–	–	–		–	52.25	1.37	5.96	40.42
[c]AN	10	–	–	–		–	62.79	1.83	2.64	32.74
[a]AN600	3723	0.81	1.85	44	56	1.6	50.79	0.22	3.95	45.04
[a]AN700	2765	0.38	1.11	34	66	3.1	59.30	0.22	3.28	37.20
[a]AN800	2481	0.35	1.31	26	74	4.2	67.11	0.15	3.63	29.11
[a]AAN	1193	0.31	0.87	36	64	2.9	65.72	0.83	3.57	29.88

1. [a] Surface area was calculated using the Brunauer – Emmett – Teller method at $P/P_0 = 0.01–1$
2. [b]V_{micro} was calculated according to a *t*-plot analysis
3. [c]V_{total} was measured at a related pressure (P/P_o) of 0.99

Reproduced with permission from Springer Nature

and 40 nm and high surface area of 42.1 m^2/g and pore size 1.9 nm. When used as an electrode, the carbon nanospheres showed high sensitivity toward nitrates with detection limit of 0.015 μM (Yallappa 2018). Areca nut carbon obtained through pyrolysis was combined with palladium to develop a catalyst for reduction of nitroarene. Carbon nanospheres obtained had surface area of 143 m^2/g, but reduced to 18 m^2/g due to the deposition of the Pd particles inside the carbon nanospheres. The catalyst developed provided better stability, catalytic loading, reaction time, and lower cost compared to metal-based catalysts (Supriya 2020). Similarly, Pd nanostructures were dispersed onto areca nut-derived carbon nanospheres (surface area of 143 m^2/g, total pore volume of 0.0658 cm^3/g) and for TEMPO mediated electrooxidation of pyridyl carbinol (Fig. 2.35). The electrode developed was able to attract pyridyl carbinol, which eventually diffused into the porous carbon structure with considerably higher efficiency than bare carbon electrodes (Mathew 2020).

In a different approach, carbon nanospheres prepared from areca nut through pyrolysis at 550 °C were combined with epoxy resin in 0.05, 0.1, 0.25, and 0.5 wt% with an intention to improve mechanical properties. Marginal improvement in tensile strength but considerable increase in elongation and modulus was observed due to the addition of the carbon (Fig. 2.36) (Hegde 2024).

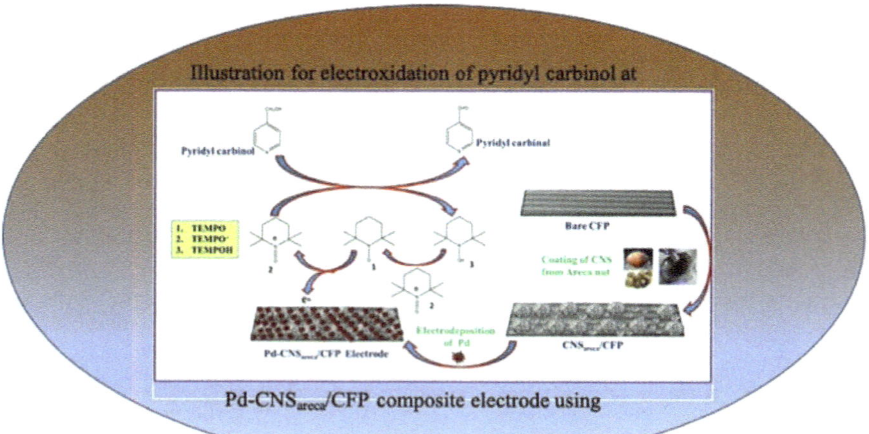

Fig. 2.35 Schematic illustration for electrooxidation of pyridyl carbinol at Pd-CNS areca/CFP composite electrode using TEMPO as a mediator (Mathew 2020). Reproduced with permission from Elsevier

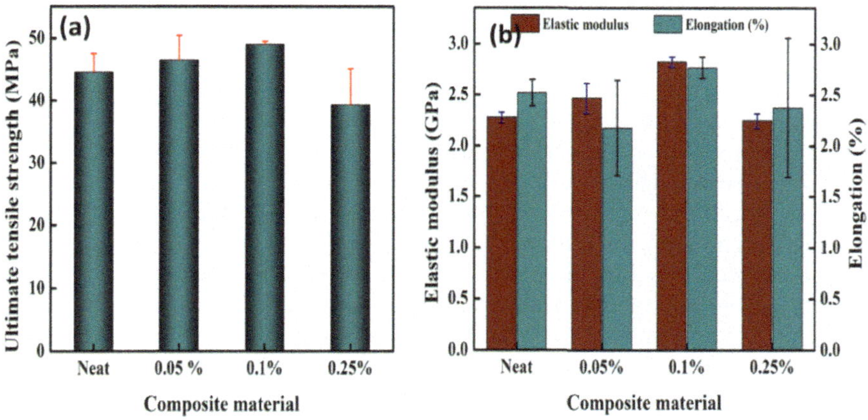

Fig. 2.36 **a** Tensile strength of composites, **b** Elastic modulus and Elongation (%) of composites compared with neat epoxy (Hegde 2024). Reproduced with permission through open access publishing

2.4.4 Miscellaneous Applications

Aqueous extracts from areca nuts were studied for their potential to dye cotton fabric. The natural dye imparted unique colors without the need for any mordants. However, mordants increased color strength and hue with K/S values ranging from 1.9 to 6.4. Excellent fastness to washing, rubbing, and light was obtained when mordants were applied (Wijayarathna 2023).

Areca nut extracts have shown the potential to be used as a dye sensitizer with desirable photochemical characteristics. To obtain the dye sensitizer, the nuts were treated with methanol, ethanol, chloroform, acetonitrile, and hexane, and the extracted solution was added to chenodeoxycholic acid (CDCA) to stabilize the solution. Sorption of the dye was highly dependent on the solvent used, with strong absorption noticed in the UV-Vis region, with maximum absorption in hexane and in methanol among the organic solvents used. Similarly, optimum performance of the cell was obtained when 1.5 mM of CDCA was used. A maximum photo conversion efficiency of 0.123% with short-circuit current density (J_{sc}) = 0.3 mA/cm^2, open circuit voltage (V_{oc}) = 0.536 V, and fill factor (FF) = 73.5% were obtained, suggesting suitability for developing a natural dye-based dye-sensitized solar cells (Najm 2020).

2.5 Conclusions

Areca nuts have been infamous and tarnished for their cancer-causing effects. Alkalioids, particularly, arecoline, the major constituent in areca nuts, have been considered the primary cause for carcinogenesis. Recent studies have, however, demonstrated that several additional components are also responsible for the ill effects of areca consumption. In addition to cancer, areca nuts have been attributed to causing inflammatory diseases and also affect fertility. However, areca nut components have been found to have several medicinal benefits, including antimicrobial and antioxidant activities, and are even able to improve cognitive ability in patients with dementia. Apart from medical applications, areca nut components have been used for synthesis of nanoparticles, to remove pollutants from the environment, and also for developing activated carbon for energy storage applications. Areca nuts vary considerably in composition, structure, and texture depending on the variety, geographical area, and conditions during growth. Comprehensive studies to compare the properties of areca nuts from different sources may lead to identifying varieties that are less toxic and more beneficial to human health.

Chapter 3
Areca Husks and Their Industrial Applications

3.1 Introduction

Being a sustainable and renewable resource with large availability, low cost, and unique chemical and physical properties, areca husks are ideal materials preferred for various applications. Areca husks are the outer covering of the nuts and account for about 50–60% of the weight of the whole fruit. Husks are inevitably removed (dehusked) and considered byproducts and are typically disposed of as waste. The husks are lignocellulosic materials like most agricultural residues. Holocellulose typically is the major component with about 55% share, followed by hemicellulose (30%) and lignin (25–30%). Among the carbohydrates (54%), glucose 34%, xylose, 18%, galactose, 0.9% are the major components. Other major constituents of areca husks are given in Table 3.1 (Yuan 2023). The lignin content of 28–30% in the husk is considerably high and is similar to that reported in coconut husks and coir fibers. However, the composition of the areca husk fibers varies considerably depending on the species, growing conditions, and extraction methods used (Sunny 2021; Banagar 2024). In terms of chemical structure, the husks show the presence of C–H and O–H vibrations, C–H bond stretching, and asymmetric vibration of CH_2 in cellulose, lignin, and hemicellulose. Further, FTIR peaks observed at 2348 cm^{-1} confirmed the presence of H–C=O aldehyde stretching and a peak at 1722 cm^{-1} due to the C=O stretching of ketones (Vardhan 2020) (Table 3.2). Phytochemical analysis has shown that the husks consist of alkaloids, steroids, phenols, and tannins with a range between 290–558 mg GAE/g depending on the type of solvent used for the extraction of the phenolic compounds (Sapaat 2022). Similarly, the lignin in the husks is composed of aliphatic OH, Guaiacyl OH, and other groups as shown in Table 3.3. Some of the major applications of areca husks are elucidated below.

N. Reddy and P. Aramwit, *Areca catechu and its byproducts*,
https://doi.org/10.1007/978-3-032-22134-6_3

Table 3.1 Chemical composition of the areca husks (Yuan 2023)

Chemical composition, %	Areca nut husk
Klason lignin	26.56
Acid-soluble lignin	2.19
Total lignin content	28.75
Holocellulose	56.10
Glucose	35.00
Xylose	18.29
Galactose	0.99
Arabinose	1.41
Mannose	0.34
Total carbohydrate content	56.03
Benzene alcohol extractive	10.23
Hot water extractive	21.33
1% NaOH extractive	40.92
Ash	3.53

Reproduced with permission through open access publication

Table 3.2 Functional groups present in areca husk lignin (Yuan 2023)

Functional group contents (mmol/g)	AHMWL
Aliphatic OH	2.77
5-Substituted OH	0.04
Guaiacyl OH	0.62
p-Hydroxyphenyl OH	0.85
Total phenolic OH	1.51
Carboxyl	0.02

Reproduced with permission through open access publishing

Table 3.3 Major peaks observed from the FTIR spectrum of areca husk fibers and their tentative assignments (Vardhan 2020)

Absorption frequency, cm^{-1}	Tentative assignments
3253	C–H, O–H stretching vibration
2922	−C–H stretching in alkane and alkyl group
2348	–H–C=O stretching in aldehyde group
1722	C=O stretching in ketones
1509	Aromatic skeletal vibration and methoxyl groups of lignin
1243	Stretching vibration of C–O group of esters and phenols

3.2 Sorption of Pollutants

3.2.1 Sorption of Dyes Using Carbon/Activated Carbon from Areca Husks

Biomass-based resources are considered ideal for the removal of pollutants due to their biodegradability, low cost, and availability in large quantities. Although areca husks have been found to be suitable for sorption of dyes and other pollutants, the presence of a hard outer layer and non-cellulosic content inhibits sorption. To increase the sorption of dyes by areca husks, they were made into particles with a size between 300 and 500 μm and treated with 0.1 M NaOH for 2 h, and their ability to sorb brilliant green dye was explored (Baidya 2021). Surface treatment removed non-cellulosic content and increased pore sizes and accessibility of the pores (Fig. 3.1). It was also observed that these pores were filled up as the dye was sorbed onto the fibers (Table 3.4). The sorption conditions, such as pH, temperature, and initial dye concentration, were also found to considerably affect the sorption potential. Based on Langmuir isotherms, the maximum sorption was found to be 18.2 mg/g with 97% dye removal, and the process was considered to be endothermic. Optimum sorption occurred at pH 7, adsorbent concentration of 10 g/L, initial dye concentration of 100 mg/L and sorption time of 120 min (Baidya 2021). Treating areca husks with sodium hydroxide was also found to be more suitable than treating with phosphoric acid, sodium carbonate, or dipotassium phosphate for sorption of basic red dye from wastewater. A spontaneous endothermic sorption of the dye with an activation energy of 10.23 kJ/mol was reported, and the sorption process followed a pseudo-second order kinetics model. A maximum sorption of 23.9 mg/g of dye was possible with a removal efficiency of 97% (Baidya 2020). Treating areca husks with alkali caused substantial chemical and physical changes but increased the ability to sorb methylene blue from wastewater (Novera 2021). Areca husks were treated at 1:1 to 1:3 weight ratio of husks to alkali at room temperature for 24 h, and the treated husks were washed, dried, and used for sorption of the dyes. Alkali treatment increased carbon content from 18 to 48%, and BET surface area showed a marginal increase from 0.916 to 0.948 m^2/g. Amount of dye sorbed by the modified husks was highly dependent on the conditions, such as temperature and initial dye and sorbent concentration (Table 3.5). Up to 145 mg/g of dye could be sorbed from the solution, and excellent desorption of 93% could be obtained by treating in salt water (Novera 2021).

Areca husks have been extensively used as a source of carbon intended for various end uses. In one such study, areca husks were treated with sulfuric acid in 1:1 ratio and later heated at 550°C for 1 h. The carbon obtained was powdered to different particle sizes and studied for their potential to sorb crystal violet dye. Sorption of the dye onto the carbon was dependent not only on the properties of the carbon but also on the conditions used. A maximum dye sorption of 76.9 mg/g was obtained, which was considered to be higher than that reported for many other biomass carbon-based sorbents (Basker 2014). A substantially higher sorption of up to 381.6 mg/g of methylene blue was possible when areca husks were made into activated carbon

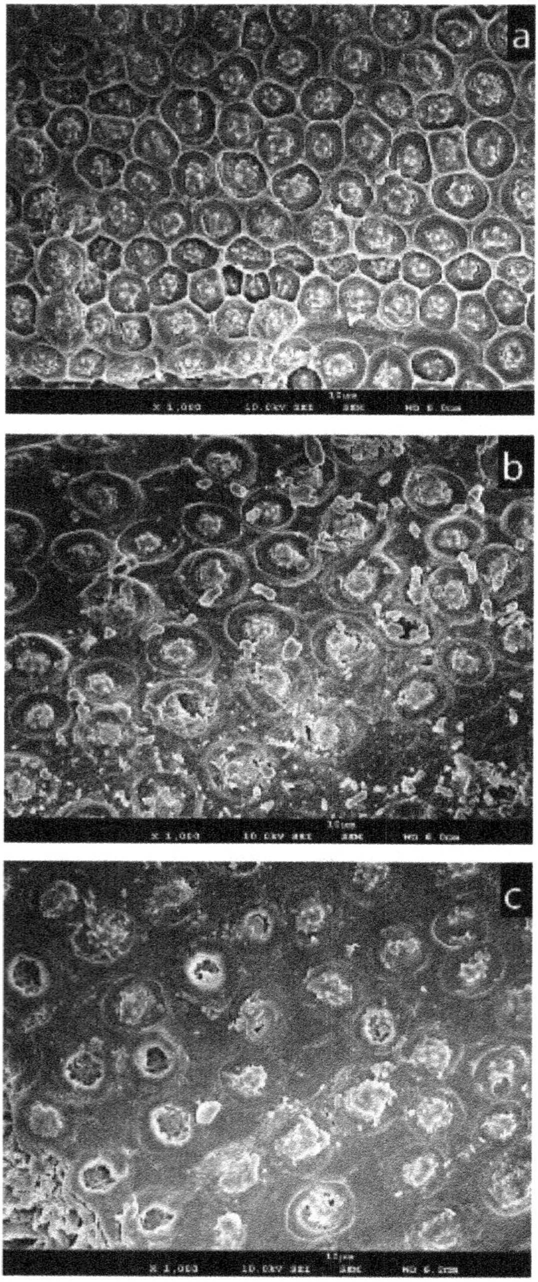

Fig. 3.1 Changes in the morphology of the areca husks before treatment (**a**), after treating with sodium hydroxide (**b**), and after sorption of the dye (**c**) (Baidya 2021). Reproduced with permission from Elsevier through open access publishing

Table 3.4 Some of the parameters studied and their influence on sorption of brilliant green dye onto sodium hydroxide-treated areca husks (Baidya 2021)

Absorbent, mg/L	Sorption, mg/g	Pseudo-first-order model			Pseudo-second-order model		
		mg/g	K_1	R^2	mg/g	K_1	R^2
50	4.832	1.613	0.037	0.792	4.919	0.071	0.999
100	9.622	5.453	0.041	0.930	10.152	0.015	0.999
200	18.314	10.812	0.041	0.942	19.380	0.007	0.999
298 K	4.832	1.613	0.037	0.792	4.919	0.071	0.999
308 K	4.847	1.537	0.0396	0.812	4.929	0.081	0.999
318 K	4.853	1.466	0.0386	0.802	4.929	0.083	0.999

Table 3.5 Kinetic parameters for the sorption of methylene blue by alkali-treated arecanut husks at different initial dye concentrations (Novera 2021). Average relative error function (ARE) and normalized standard deviation (NSD)

Parameter	Initial dye concentration (mg/L)					
	25	50	100	150	200	250
Q_{eexp} (mg/g)	23.77	45.38	91.99	118.95	131.08	162.6
Pseudo-first order						
Qc (mg/g)	22.44	44.07	85.0	113.03	126.69	166.74
K_1 (l/min)	0.02	0.06	0.004	0.025	0.02	0.014
R^2	0.894	0.936	0.998	0.741	0.815	0.831
NRD	15.34	7.425	18.038	3.685	2.306	14.465
ARE	1.66	0.023	0.001	0.486	4.374	0.053
Pseudo-second order						
Qc (mg/g)	24.491	45.62	94.44	211.86	229.95	263.93
K_1 (l/min)	0.003	0.003	0.0007	0.00002	0.00002	0.00003
R^2	0.981	0.892	0.975	0.659	0.572	0.467
NRD	6.633	18.951	23.056	29.613	31.004	21.104
ARE	0.008	0.007	0.002	5.787	9.356	0.2

after treating with sodium hydroxide and heating up to 500 °C for 1 h under nitrogen atmosphere. The carbon obtained had a surface area of 81.7 m^2/g and total pore volume of 0.722 cm^3/g, and mean pore diameter of 3.53 nm making the material mesoporous. Extent of sorption was dependent on sorption conditions, particularly temperature with optimum being at 30 °C and pH 4. Dye sorbed onto the carbon could be easily desorbed by treating in saline water suggesting that the carbon could be recycled (Bardhan 2020) (Fig. 3.2). Another study reported that carbon prepared from areca husks had surface area of 310 m^2/g and was found to be suitable for sorption of methylene blue (Basker 2015). Areca husks were carbonized at 550 °C, and the carbon obtained was separated into three (100, 150, and 250 mesh) sizes and used for sorption of malachite green (Basker 2014). As reported in other studies, the sorption was found to be exothermic and was highly dependent on the sorption temperature, with enthalpy of −23 kJ/mol at 20 mg/l dye concentration and −16.7 kJ/mol at 40 mg/l. Sorption followed the Langmuir isotherm, and under the optimized conditions, a maximum sorption of 59.2 mg/g was obtained (Basker 2014, 2015). Activated carbon obtained from areca husks having surface area of about 310 m^2/g was found suitable for sorption of methylene blue. Optimum sorption of dye was between pH 9–10 with 70% removal efficiency, which is considerably lower than that reported by most other studies using activated carbon from areca husks and other biomass sources. Such lower absorption is probably due to the unoptimized conditions and also nature of carbon prepared. Areca husks were subject to hydrothermal heating at 200 °C for 1 h, and the biochar obtained was subsequently chemically activated by combining with NaOH at 1:2 ratio and heating the mixture at 500 °C for 1 h. Carbon (biochar) obtained had surface area of 518 m^2/g, pore diameter of 25 nm, and pore volume of 0.033 cm^3/g. When used for sorption of methylene blue, sorption capacity ranged between 324 and 430 mg/g depending on the temperature during sorption in an endothermic process (Tabassum 2020). In a similar study, areca husks were made into biochar by heating between 350 and 500 °C under nitrogen atmosphere and used for dye removal. Biochar obtained showed high but non-uniform pores and pore tunnels, which were suggested to be useful for sorption of Rhodamine B (RhB) and anionic Alizarine red S (AZ) dyes. An adsorption efficiency of 89 for blue and 87% for the red dye was obtained with adsorption capacities of 105 mg/g and 109 mg/g, respectively, higher than most other biochars developed previously for dye removal (Tharayil 2023).

To enhance the capability of activated carbon to remove methylene blue, activated carbon from areca husks was grafted with CeO_2 quantum dots (QDts) using *Saccharum officinarum* extract as schematically shown in Fig. 3.3. X-ray diffraction analysis showed that the biochar was amorphous in nature and that the CeO_2 particles were embedded into the matrix without any distinct phase separation. Addition of CeO_2 provided the nanocomposites considerably higher surface area and higher active sites. Specific surface area of the activated carbon was 40 m^2/g with pore diameter of 1.98 nm and pore volume of 0.81 cc/g. A considerably high absorption capacity of 492 mg/g of methylene blue was achieved mainly through electrostatic attraction, pi–pi interaction, etc. The composite also had high recyclability with absorption efficiency as high as 93.7% even after five cycles (Gupta 2024).

Using a similar approach, areca husks were converted into hydrochar and nanohydrochar for potential use as sorbents for Congo red in waste water. Hydrochar was prepared by heating the husks in a hydrothermal autoclave at 250 °C for 4 h. Resulting solid material was collected and washed thoroughly to remove impurities and taken as the absorbent. To prepare the nano hydrochar, the hydrochar obtained was placed in a high-energy milling machine and operated at 500 rpm for 3 h. Resulting material was sieved through a 200 mesh filter and stored for further use. X-ray diffraction analysis showed that the nanochar had higher crystallinity and oriented structure than

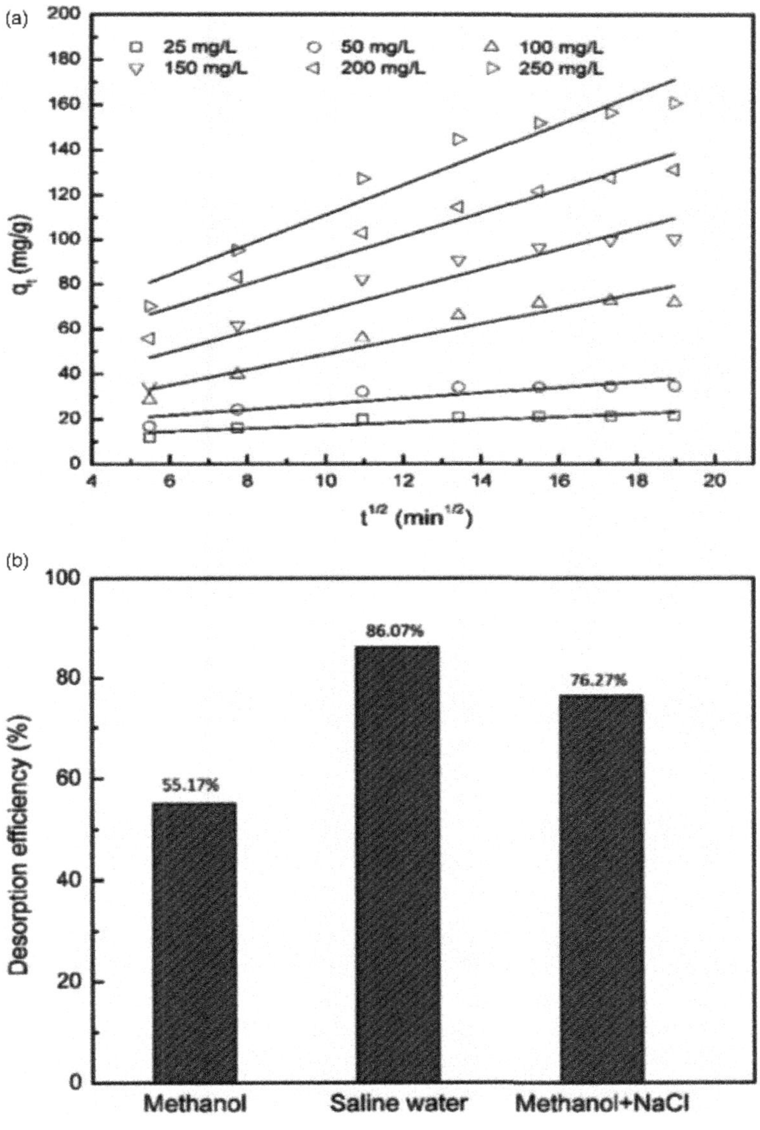

Fig. 3.2 Sorption kinetics and level of desorption of methylene blue by the activated carbon prepared from areca husks (Bardhan 2020). Reproduced with permission through Creative Commons Attribution Licence (CC BY 4.0)

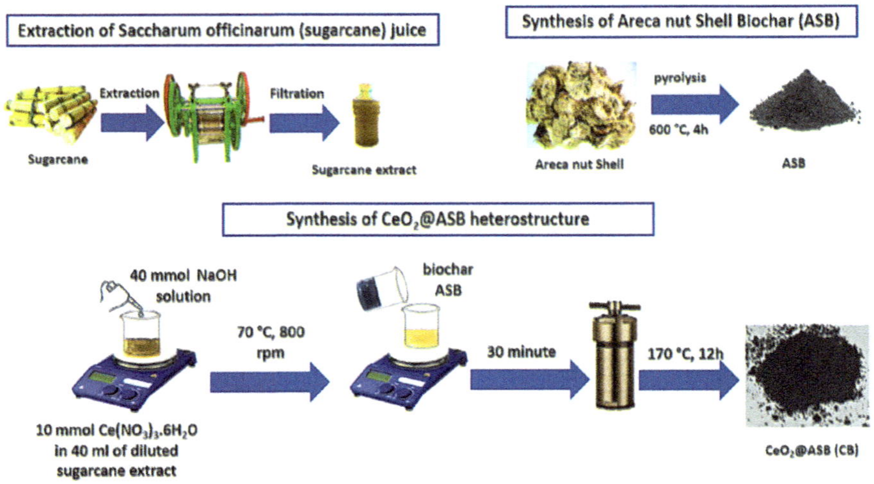

Fig. 3.3 Schematic representation of the process used to obtain the areca husk carbon and subsequent grafting of CeO_2 quantum dots (Gupta 2024). Reproduced with permission from Elsevier

the biochar. Nanochar had higher surface area, smaller particles, smoother surface with several pores that would assist in dye sorption (Fig. 3.4). Hence, the nanochar had the highest sorption capacity of 596 mg/g compared to 356 mg/g for the biochar. Sorption followed pseudo-second order kinetics with isotherm models indicating chemisorption and monolayer adsorption (Adawiyah 2025).

3.2.2 Removal of Metal Pollutants Using Areca Husks

An attempt was made to produce biochar from areca husks on a commercial scale using a drum kiln setup (Fig. 3.5) (Mitra 2024) for potential use as a sorbent for heavy metals. In this process, areca husks were heated in the kiln at temperatures between 350 and 500 °C under limited oxygen to form the biochar. The biochar formed was highly porous with macropores having size between 5 and 20 μm. Sorption capacity of the biochar depended on the conditions, with maximum sorption of 1.73 and 1.77 mg/g for Ni^{2+} and Mn^{2+} at pH 7. Such high absorption was possible due to the electrostatic interaction between the biochar surface and metal ions, since the surface of the biochar is negatively charged, whereas the metal ions contain positive charges. Equilibrium sorption was reached at 150 min of sorption for Ni^{2+} and at 180 min for Mn^{2+}. Kinetic studies (Table 3.6) showed that the sorption followed the pseudo-second-order model and the Freundlich isotherm (Mitra 2024).

In another study, carbon was prepared by heating powdered areca shells at 800 °C for 3 h. Subsequently, the carbon obtained was treated with sodium hydroxide at 27 °C for 3 h and later washed with distilled water and dried. Although the properties of the

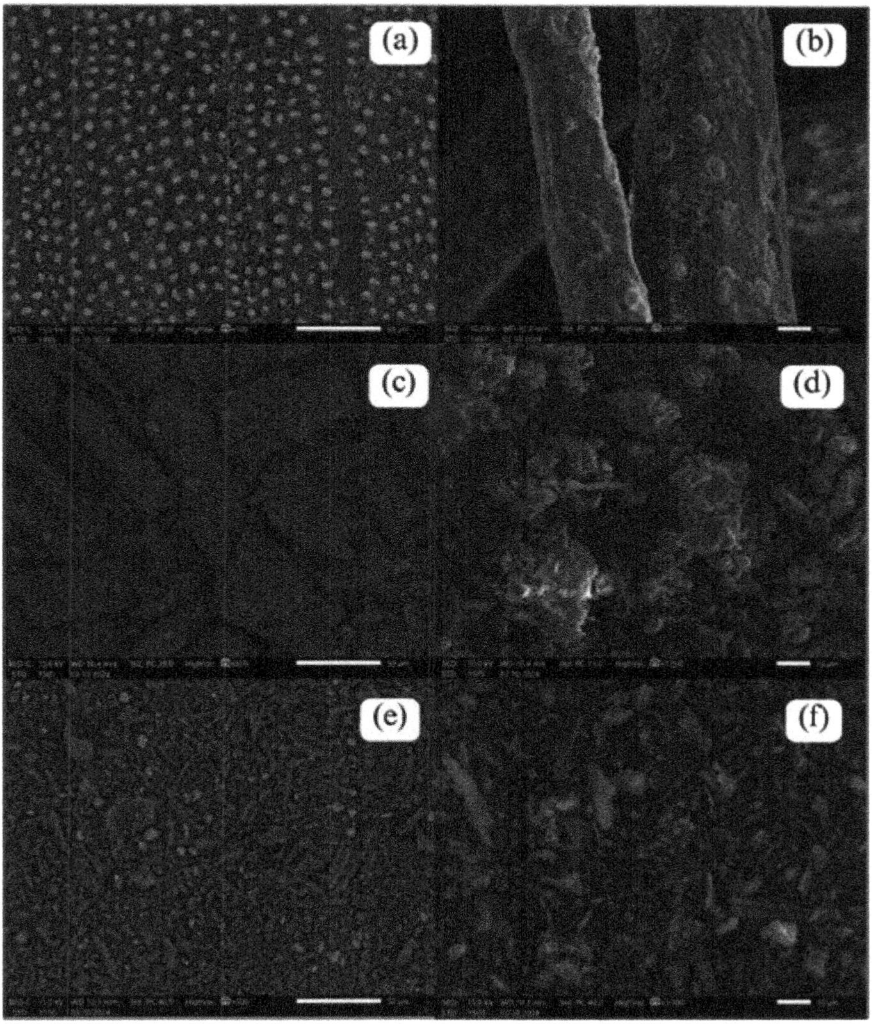

Fig. 3.4 SEM image of Areca husk x500 (**a**), Areca husk x1000 (**b**), HC x500 (**c**), HC x1000 (**d**), nano HC x500 (**e**), and nano HC x1000 (**f**) (Adawiyah 2025). Reproduced with permission through open access publishing

carbon obtained were not reported, it was found that the activated carbon could sorb Pb(II) ions from aqueous solutions through pseudo-second order kinetics, and the amount of ions sorbed was dependent on temperature, pH, etc. A sorption of 50.5 mg/g at 27 °C and 55.3 mg/g at 45 °C was obtained. It was suggested that the sorption process was endothermic and that both physical and chemical adsorption should have occurred (Muslim 2017). In another study, biochar prepared by pyrolyzing areca husks at 450 °C for 2 h was studied for its potential to remove Fe^{2+} ions in aqueous

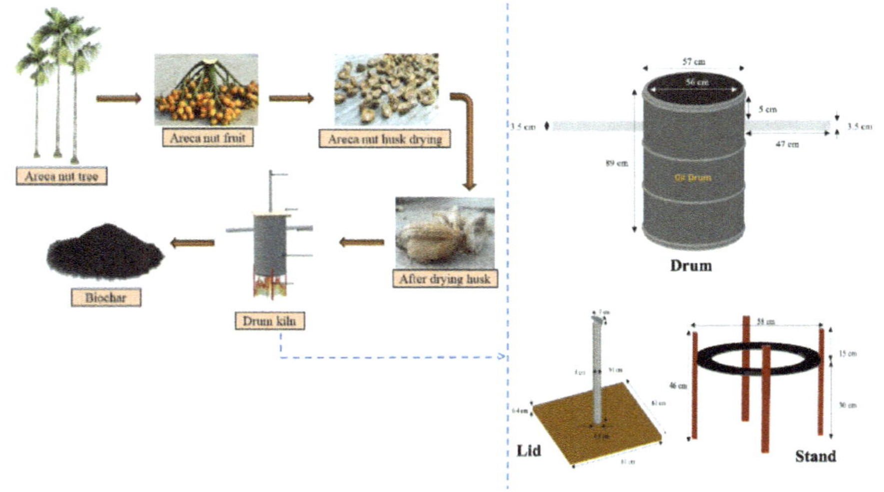

Fig. 3.5 Drum kiln setup for mass-scale biochar production (Mitra 2024). Reproduced with permission from Springer Nature

Table 3.6 Some of the kinetic parameters for sorption of Ni^{2+} and Mn^{2+} ions by areca nut biochar (Mitra 2024)

		PFO				PSO		
Metal	Initial Conc (mg L^{-1})	Qe,exp (mg g^{-1})	KPFO	Qe,cal (mg g^{-1})	R^2	KPSO	Qe,cal (mg g^{-1})	R^2
Nickel	10	1.140	0.013	0.212	0.623	0.340	1.167	0.999
	15	1.389	0.013	0.497	0.913	0.063	1.473	0.998
	20	1.563	0.014	0.726	0.985	0.043	1.760	0.998
	25	1.651	0.013	0.848	0.967	0.031	1.919	0.997
	30	1.727	0.017	0.995	0.801	0.044	1.960	0.997
Manganese	10	1.051	0.004	0.075	0.108	0.030	1.389	0.969
	15	1.007	0.004	0.149	0.098	0.084	1.815	0.937
	20	1.555	0.003	0.272	0.077	0.064	1.818	0.990
	25	1.638	0.005	1.023	0.585	0.005	2.159	0.804
	30	1.782	0.005	0.406	0.108	0.022	1.24	0.826

Reproduced with permission from Springer Nature

solutions (Subramani 2019). Biochar obtained had about 75% carbon content, and the surface had acidic functional groups preferable for sorption of heavy metal ions. The biochar had a specific surface area of 54.5 m^2/g and pore volume of 0.0142 cm^3/g. In addition to these favorable properties suitable for sorption, it was estimated that the arecanut husk-based biochar could be obtained at $1.4/kg, which was considerably lower than charcoal or other biobased biochars. During optimization studies, it was found that monolayer adsorption, physisorption, and chemisorption occurred mainly through pseudo-second order kinetics. Maximum sorption of Fe^{2+} ions was 0.38 mg/g, and the process was spontaneous and exothermic (Subramani 2019). Further, up to 95% of the metal ions could be desorbed in 0.1 M HNO_3, indicating easy recyclability.

In a different approach, impregnation of activated carbon obtained from areca husks with iron was done to improve the sorption of Cu(II) and Pb(II). To obtain the modified carbon, activated carbon was treated with 0.025 M iron nitrate solutions having pH 9 at 60 °C and heated at 110 °C for 2 h. About 107 mg/kg of iron before modification and 2442 mg/g of iron after modification were found on the carbon, mostly distributed on the surface. No major change in the BET surface area of the carbon was observed either before or after modification. Under the optimized conditions, the sorption capacity of the carbon was 1.926 mg/g for Cu(II) and 3.439 mg/g for Pb(II). Impregnation with iron was found to substantially improve the sorption capacity for both the pollutants (Tiwari 2015). Activated carbon prepared from areca husks was also found to be suitable for removal of Cr(VI), and the extent of removal was based on the initial concentration, and up to 45% chromium could be removed even at a low carbon concentration of 0.06 mg/ml (Renugadevi 2010).

3.2.3 Removal of Phenol

Areca husks have also shown potential to degrade phenol when modified with microorganisms. For the study, areca husks were washed, dried, and then cut into pieces of 5–6 mm. These treated husks were used for immobilization of microorganisms by natural adsorption. Bacterial culture used was 0.944 g/L of dry weight and 0.866 g/L of dry cell biomass, which was suspended in phenol. Conditions such as pH, incubation temperature, and initial phenol concentration were varied, and the extent of degradation of phenol was determined. Morphological images showed large number of bacterial deposits on the surface and, in some cases, even film formation. Substantial degradation of phenol from an initial concentration of about 200 mg/L was possible depending on the conditions used. An optimum pH of 8 and temperature of 37 °C was capable of degrading 1000 mg/L of phenol in suspension cultures in 28 h with highest degradation rate of 0.37 g per hour. The enzymes immobilized on the husks could be stored for several weeks without any major decrease in efficiency (Bera 2020). In a study by Girish and George, it was found that areca husks treated with phosphoric acid (1:4 ratio) had a considerably high surface area of $679m^2/g$ and could sorb 27.7 mg/g of phenol under the optimized conditions (Girish 2017). Activated carbon prepared from areca husks using phosphoric acid as the activation agent was found to be suitable for removal of Bisphenol-A (BPA). Activation was done at 550 °C for 2 h, and carbon formed had a high surface area of 483 m^2/g and pore volume of 0.87 cm^3/g, which provided a high adsorption potential of 28.4 mg of bisphenol per gram of the activated carbon. The carbon developed was hence considered suitable to treat water polluted with BPA, particularly in the plastic

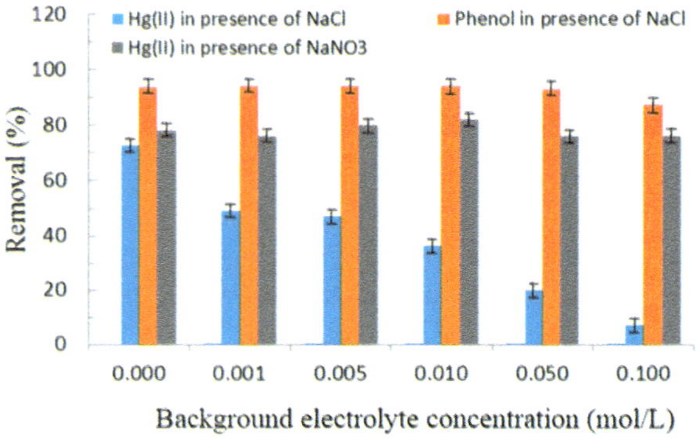

Fig. 3.6 Effect of background electrolytes on the simultaneous removal of phenol and Hg(II) by succinic anhydride modified activated carbon prepared from areca husks (Lalhmunsiama 2017). Reproduced with permission through open access publication

manufacturing industries (Yong 2024). Activated carbon prepared from areca husks and modified by treating with succinic anhydride was studied for its potential to simultaneously remove mercury (Hg(II)) and phenol under batch and fixed column systems (Lalhmunsiama 2017). SEM images showed that the activated carbon was highly porous, and the pore diameters were determined to be between 5 and 15 μm. Surface area of the carbon was 367 m^2/g, which decreased to 308 m^2/g after treating with succinic anhydride. Similarly, pore size and pore volume changed from 3.13 to 2.78 nm and 0.028 to 0.016 cm^3/g, respectively after modification of the carbon. Ability of the unmodified and modified activated carbons showed that the level of sorption of phenol and Hg (II) was dependent on conditions such as pH, initial concentration, etc. It was found that increasing pH increased sorption of Hg(II) but not phenol. Maximum sorption capacity was 11.23 mg/g for Hg(II) and 5.4 mg/g for phenol, and loading capacity was 2.5 mg/g and 2.7 mg/g for mercury and phenol, respectively. Presence of background electrolytes such as NaCl and $NaNO_3$ also affected the extent of removal of the pollutants (Fig. 3.6) (Lalhmunsiama 2017). Other than phenol and mercury, areca husk carbon activated using H_3PO_4 at 500 °C for 120 min resulted in carbon with surface area between 580 and 780 m^2/g and absorption capacity of up to 67 mg/g for paracetamol and various other pharmaceuticals primarily through Langmuir monolayer adsorption (Mandal 2024). It was also possible to regenerate and reuse the absorbent for up to five cycles without major reduction in sorption potential.

3.3 Energy Applications

3.3.1 Activated Carbon and Supercapacitors

A one-step carbonization and activation was used to obtain nanofibrous carbon from areca husks with excellent surface properties suitable for energy storage applications. In this study, fibers were separated from areca husks and powdered into 53 μm particles. This powder was treated with 0.3–0.9 M KOH and heated initially at 600 °C and later at 900 °C for 2.5 h under nitrogen atmosphere to convert into carbon. The morphology and physical properties of the carbon obtained varied significantly depending on the amount of KOH used during activation (Taer 2020) (Fig. 3.7). The carbon was used to develop a symmetrical supercapacitor, and the cyclic performance of the device was measured with 1 M sulfuric acid as the electrolyte. Carbon obtained under the optimized conditions provided a large surface area of 839 m^2/g and total pore volume of 0.448 cm^3/g. Due to the unique structure of the carbon, a high specific capacitance of 182 F/g and maximum energy density of 25.3 Wh/kg and power density of 91.7 W/kg were obtained (Table 3.7). It was suggested that the nanofibrous structure promoted charge transfer and increased access to the electrolyte, leading to better supercapacitor performance (Taer 2020). In another study, areca husks were treated with KOH and later carbonized by heating at 900 °C from 1.5 to 3.0 h. Carbon obtained was used as an electrode and the electrochemical properties were studied for potential use in supercapacitor applications (Taer 2019). Interestingly, the carbon obtained was in fibrous form and resembled carbon nanotubes. Particle size of the carbon was between 698 and 3313 nm with corresponding length between 1034 and 4852 nm under the optimized carbonization time of 2.5 h. Many of the nanofibers had a hollow structure with inner diameter between 40 and 80 nm and outer diameter between 70 and 122 nm. BET surface area of the activated carbon fibers ranged from 482 to 757 m^2/g and pore size was between 3.1 and 3.2 nm. When made into an electrode, a specific capacitance of 165 F/g was obtained with highest energy density of 5.73 Wh/kg and power density of 41.3 W/kg (Taer 2019). The same researchers reported the production of carbon nanofibers from areca husks with properties suitable for use as supercapacitors. The husks were initially carbonized and later activated with potassium hydroxide. Carbon formed had considerably high surface area of 838 m^2/g, total pore volume of 0.448 cm/g and exhibited a specific capacitance of 182 F/g, energy density of 25 Wh/kg and power density of 91 W/kg indicating that areca husks could be ideal resources for production of carbon for energy applications (Taer 2021).

Areca husks were carbonized by heating at 800 °C for 2 h to obtain a yield of about 61%. The carbon obtained was activated using KOH and studied for potential supercapacitor applications. Carbon obtained had average pore diameter of 1.69 nm, specific surface area of 1197 cm^2/g, micro and total pore volume of 998 cm^2/g (Swathy 2024). When made into supercapacitors, the carbon showed a high capacitance of 217 F/g, specific capacitance of 46 F/g, specific energy of 6.1 Wh/kg with high capacitance retention of 92% even after 7500 cycles. A flexible and binder-free supercapacitor electrode was developed using nanocellulose (20–50 nm diameter) extracted from areca husks combined with polyaniline embedded with silver nanoparticles (Sasi 2021). A specific capacitance of 780 F/g, energy density of 15.64 W h/Kg and power density of 244.8 W/kg were obtained for the electrodes developed. Inclusion of the nanocellulose was found to increase the specific capacitance by about 4.2 times (Fig. 3.8).

An attempt was made to further improve the electrochemical performance of carbon derived from areca husks. A composite was prepared by combining activated

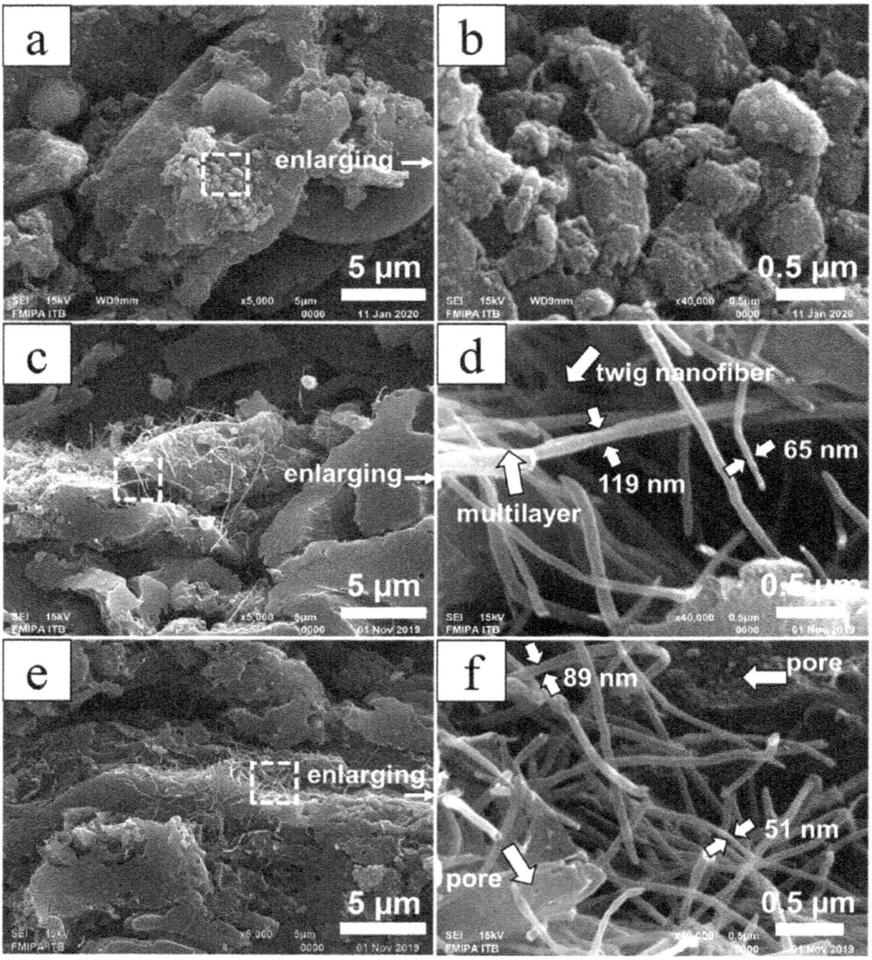

Fig. 3.7 SEM images show the surface of the unactivated carbon (**a**, **b**), 0.3 M KOH activated (**c**, **d**), and 0.7 M KOH activated carbon (**e**, **f**) at two different magnifications distinctly show the formation of nanofibers due to KOH activation (Taer 2020). Reproduced with permission from Springer Nature

carbon from areca husk with niobium oxide nanoparticles and used as an electrode in energy storage device. When used in iron or vanadium redox flow batteries, the composite showed considerably enhanced electrochemical performance with columbic efficiencies of 87 and 89%, respectively. When used in a supercapacitor, a specific capacitance of 245 G/g and columbic efficiency of 81% even after 3000 cycles was obtained (Gireeshkumar 2023).

Areca nut ash prepared by open-air burning was calcined by heating up to 800 °C for 4 h, and the material obtained was used to fabricate anodes for Li ion batteries

Table 3.7 Changes in the surface properties of carbon obtained from areca nut husks before and after activation with two different concentrations of KOH (Taer 2020)

KOH concentration, M	Specific surface area, m^2/g		Pore volume cm^3/g		D (nm)
	BET	BJH	V total	V BJH	
0	696.7	35.3	0.375	0.040	2.136
0.3	759.9	41.7	0.403	0.050	2.151
0.7	838.6	23.6	0.448	0.032	2.137

Reproduced with permission from Springer Nature

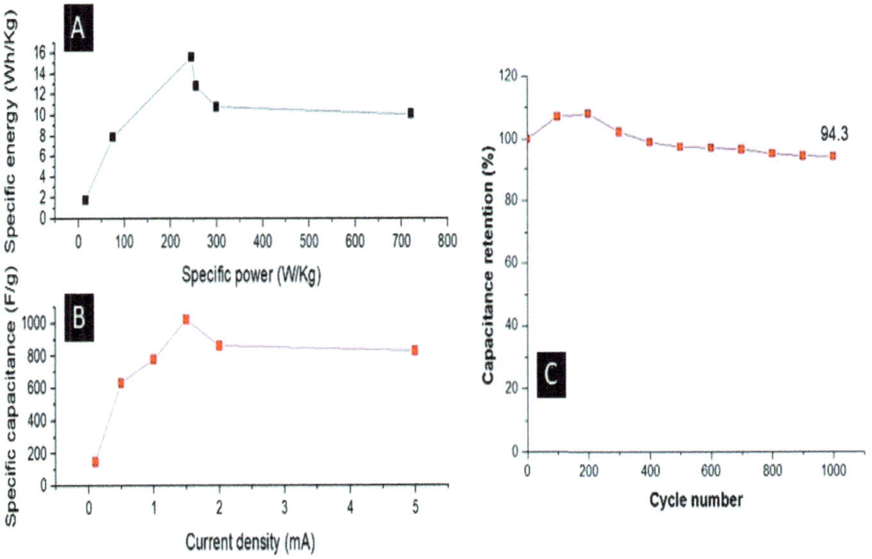

Fig. 3.8 Performance of the electrode developed using nanocellulose from areca husks and combined with polyaniline embedded silver nanoparticles (Sasi 2021). Reproduced from the Royal Society of Chemistry through Creative Commons Attribution-Non-Commercial 3.0 Unported Licence

(Panoth 2023). The resulting Si nanospheres embedded in calcium silicate matrix had highly porous structure and surface area of about 27 m^2/g. Electrochemical analysis (Fig. 3.9) showed that the areca husk ash-coated electrodes had considerably higher capacity, columbic efficiency, and current density with reversible capacity of 2109 mAh/g at 0.4 A/g, 85% capacity retention even after 100 cycles, and rate capability of 901 mAh/g at 5 A/g. In a similar approach, activated carbon prepared from areca husks was made into a composite with niobium oxide nanoparticles by heating at 200 °C for 2 h. The prepared composite was coated on an untreated graphite felt electrode for iron redox flow and vanadium redox flow batteries, and also made into an electrode for supercapacitors by combining with polyvinyl di-fluoride

(PVDF) and N-Methyl pyrrolidone and coating the mixture onto the surface of Toray carbon paper. The areca husk ash carbon-treated materials had high active surface area of 132 cm^2 and showed high columbic efficiency of 87% at a current density of 60 $mAcm^{-2}$ and 30 $mAcm^{-2}$ at a CE of 89%. Energy capacity and discharge capacity for the vanadium and iron batteries were 526 mWh and 658 mAh and 3877 mWh and 4847 mAh, respectively. In addition, the composite material showed low charge transfer resistance (0.01162 Ω), 81% columbic efficiency, and 77% specific capacitance retention even after 3000 charge discharge cycles, suggesting suitability for energy storage and energy conversion applications (Gireeshkumar 2023).

In a different approach, nanocellulose derived from areca husks was combined with silver nanoparticles and polyaniline. This composite was used as an electrode, and the electrochemical properties were investigated. A high specific capacitance of 780 F/g, energy density of 15.6 Wh/Kg, and power density of 244 W/kg were obtained with good recyclability even after deformation of the electrode to different extents (Fig. 3.10) (Sasi 2021). In another study, activated carbon from areca husks was used as a cathode catalyst in a microbial fuel cell. A relatively high surface area of 1262 m^2/g was obtained for the carbon with power density of 590 mW/m^2 and current density of 1.03 A/m^2 at 612 Ω, and hence the carbon from areca husk was considered suitable as a metal-free catalyst for fuel cells (Subran 2023).

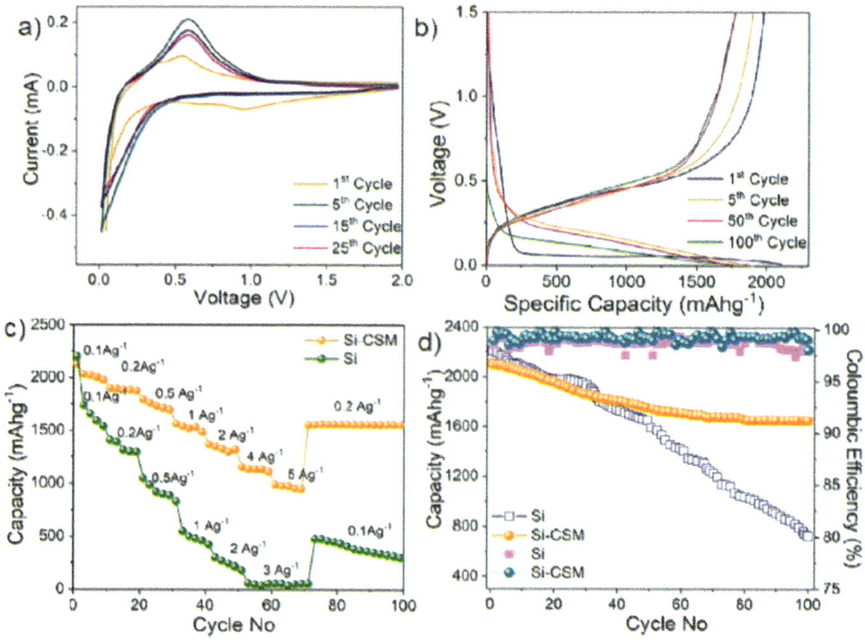

Fig. 3.9 Cyclic Voltammetry curves (**a**) and GCD curves (**b**) of Si-CSM samples. Plot showing the rate capability (**c**) and Cyclic stability (at 0.1 Ag^{-g} for Si-CSM and 0.05 Ag^{-1} for p-Si) and Columbic efficiency of Si-CSM and P-Si samples (Panoth 2023). Reproduced with permission from Elsevier

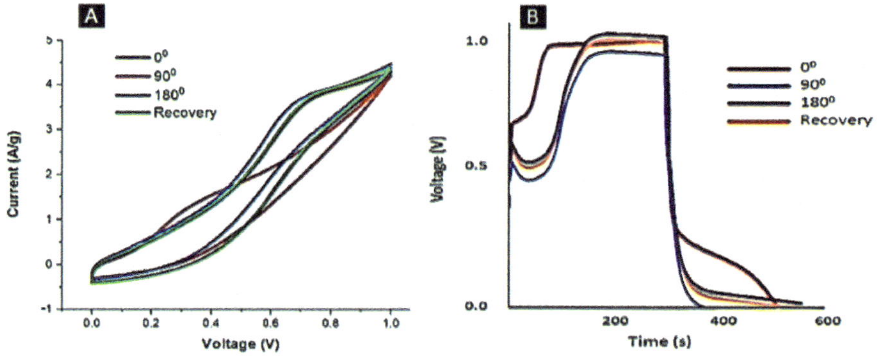

Fig. 3.10 **a** CV (A)and GCD (B), curves of the Ag–PANI–NC electrode at different scan rates and current densities, respectively (Sasi 2021). Reproduced with permission from RSC through open access publishing

3.3.2 Briquettes and Biofuels

In a direct application of using areca husks as a source of energy, the husks were combined with simarouba seed shells and made into briquettes with black liquor from pulp production as a binder. Various ratios of the raw materials were used and compressed into briquettes using pressure between 60 and 80 MPa. Density, compressive strength, composition, and calorific value of the briquettes were determined. An image of the briquette developed using four different ratios of areca nut husks and the seed shells is shown in (Fig. 3.11) (Ujinappa 2018). Some of the properties of the briquettes are provided in Table 3.8. Briquettes obtained under the optimum ratio of 60% husk and 40% seed shells showed a compressive force of 80 MPa, density of 1040 kg/m^3 and strength of 29.15 N/mm, and a calorific value of 18.8 MJ/Kg which was considered suitable for commercial applications. It has been reported that the areca nut shells available in Indonesia could be used to produce 4,653 tons of oil equivalent of renewable energy, corresponding to a CO_2 mitigation of 18,491 tons (Setyawan 2020). To utilize such an abundant bioresource, areca husks were combined with cashew shell nuts and made into briquettes using cassava flour as the binder. The three materials were combined in various ratios, and two different (small and large) size of cashew shell nuts were combined together. Briquetting was done in a fuel briquette machine operating between 70 and 90 rpm to obtain samples that were 50 mm in diameter and 100 m in length and weighed about 210 g. Up to 245 pieces per hour corresponding to a weight of about 52 kg/hour could be obtained. Calorific value of the briquettes was 33 MJ/kg, volatile matter content was 35%, fixed carbon was 14% and combustion rate was 0.62 g/min, which were similar to commercially available biochars used as fuel or wood charcoal but with considerably lower greenhouse gas emissions (Chungcharoen 2020).

A comprehensive approach was used to determine the possibility of using areca nut husks as a source for biofuels. Powdered areca husks (saw dust) were dried

Fig. 3.11 Image of the briquettes made from four different combinations of arecanut husks and simarouba seed shells and compressed at 80 MPa (Ujjinappa 2018). Reproduced with permission through Creative Commons 4.0 International License (CC-BY-SA)

Table 3.8 Properties of arecanut husks and simarouba blend (60:40) briquettes in comparison to sawdust charcoal and barley briquettes (Ujinappa 2018)

Property	Saw dust charcoal	Barley	Areca husks/simarouba seed shells
Moisture	5.70	9.55	5.75
Volatile matter	71.00	73.48	73.71
Fixed carbon	20.70	21.07	18.19
Ash	2.60	5.46	2.48
C	53.07	44.98	65.86
H	4.10	6.15	6.86
N	0.28	0.68	0.63
O	39.60	48.30	15.21
S	0.302	0.30	0.18
Calorific value (MJ/kg)	20.16	18.54	18.81

and pyrolyzed using ZSM-5 with surface area of 425 m^2/g and MgO with surface area of 195 m^2/g as the catalyst. The catalysts were combined with areca husks in 3:1 to 12:1 ratio and pyrolyzed in a semibatch reactor at temperatures between 400 and 600 °C under inert gas conditions. Pyrolysis of areca husks produced liquid, biochar, and flugas, the amount of each varied depending on the conditions (Fig. 3.12). Properties of the pyrolytic oil and biochar obtained from the areca husk are given in Tables 3.9 and 3.10. It was suggested that the catalysts provided oil with lower ash content and increased gross heating value and carbon content suitable for commercial applications (Mishra 2021). Areca husks could also be converted into pyrolytic oil when heated in a semibatch tubular reactor at 600 °C at a heating rate of 80 °C/min with and without ZSM-5 as the catalyst. The pyrolytic oil obtained was composed of OH, C–H, and C=O functional groups, ascertaining the presence of phenols, ethers, esters, alcohols, alkanes, etc. Some of the physiochemical characteristics of the oil obtained from the husks are given in Table 3.11 (Mishra 2024). In another study, the potential of pyrolyzing areca husks was studied in a thermogravimetric analyzer at heating rates between 2 and 20 °C/min. Thermal decomposition of the husks was dependent on the heating rate but was rapid above 225 °C (Fig. 3.13) with weight loss of about 44%. Average activation energy for the pyrolysis was 230 kJ/mol (Gokul 2019).

Similar to pyrolysis, the properties of biooil and biochar obtained after torrefaction of areca husks were studied by Gogoi et al. (2017). Torrefaction was done in a fixed bed system at temperature between 200 and 275 °C at a heating rate of 40 °C/min for 30 min under nitrogen atmosphere and the torrefied samples were collected. Pyrolysis of the husks was also done by heating 10 g of the material in a fixed bed tubular reactor at 300–600 °C at 40 °C/min or 20 min under nitrogen atmosphere. Biooil, biochar, and gases obtained from the pyrolysis process were collected and analyzed for their properties. Depending on the temperature used during torrefaction, the mass yield varied between 71 and 58% with energy yield between 77 and 83% and mass energy density being 1.17–1.32. Similarly, the biooil obtained from

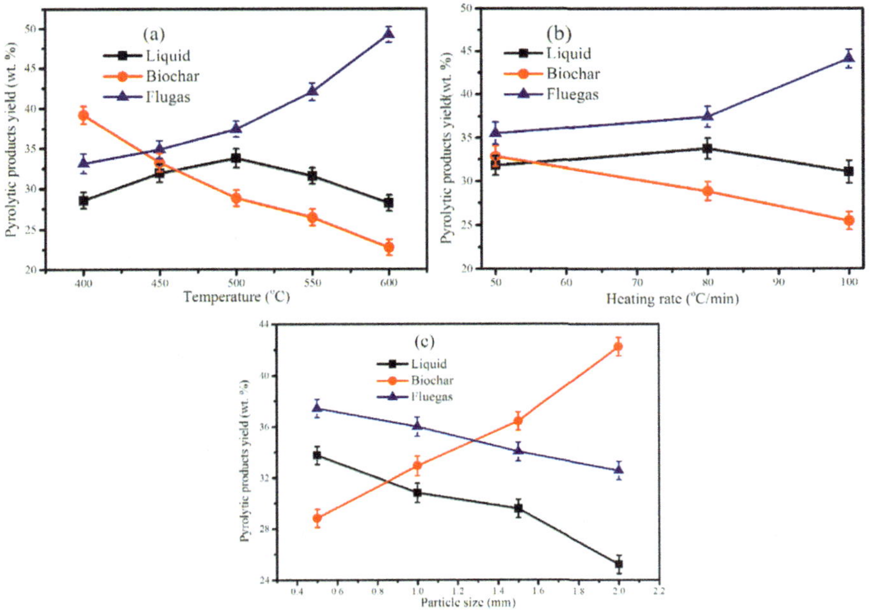

Fig. 3.12 Influence of temperature, heating rate, and particle size on the yield of liquid, biochar, and flugas from the pyrolysis of areca husk saw dust (Mishra 2021). Reproduced with permission from Elsevier

Table 3.9 Comparison of the properties of pyrolytic oil obtained from areca husks without (thermal) and with the two catalysts in comparison to commercial biodiesel (Mishra 2021)

Parameter	Thermal	MgO	ZSM-5	Diesel
Color	Black	Black	Black	–
C, %	65.12 ± 0.2	81.19 ± 0.18	76.15 ± 0.15	86.60
H,%	7.12 ± 0.1	7.19 ± 0.1	8.16 ± 0.1	13.60
O,%	25.14 ± 0.8	10.72 ± 0.4	13.59 ± 0.4	1.80
N,%	2.43 ± 0.14	1.23 ± 0.1	1.28 ± 0.1	0.0065
S,%	0.92 ± 0.01	0.72 ± 0.02	0.49 ± 0.014	0.11
Heating value (MJ/kg)	28.38 ± 1.12	38.02 ± 1.16	35.09 ± 1.14	42.43–45.50
Viscosity (cSt at 40 °C)	58.48 ± 1.02	16.16 ± 0.14	17.12 ± 1.10	2.10
Acidity	2.87 ± 0.1	8.78 ± 0.2	5.42 ± 0.2	–
Density, kg/m^3	841.0 ± 2.14	792.12 ± 2.12	764.98 ± 2.18	828–838
Moisture, %	1.30 ± 0.60	1.64 ± 0.10	1.52 ± 1.06	<0.10
Ash content, %	0.46 ± 0.014	0.26 ± 0.001	0.29 ± 0.018	–

untreated areca husks had calorific value of 27.3 MJ/kg compared to 30.2 MJ/kg for the torrefied husks. Properties of the biochar obtained from the raw and torrefied husks are given in Table 3.12. Biochar obtained through the torrefaction route was able to completely remove As (V) from aqueous solution within 70 min (Gogoi 2017). In another approach, anaerobic retting of areca husks was done in plug flow bioreactors for up to 50 days, and the production of fiber, biogas, and compost was studied (Chanakya 2012). Optimum yield of fibers from the husks was obtained after treating for 35 days. Up to 72% pectin was degraded in 10 days and 60% cellulose was degraded in 50 days with final cellulose content of 57 and 73% of lignin remaining compared to the composition before the hydrolysis. Biogas (252 ml/g total solids after 50 days of digestion) obtained had 60% methane and 40% carbon dioxide, similar to most other biomass feedstocks. Large availability of areca husks and the ability to generate multiple products using a single fermentation process were suggested to be ideal for economic and technical considerations (Chanakya 2012).

3.3.3 Production of Enzymes/Biofuels

Areca nut husks were found to be ideal substrates and source of carbon for manufacture of endoglucanase using *Bacillus subtilis MU S1* with eventual use for production of ethanol. Endoglucanase activity of 329 U/ml was possible when arecanut husks were used as substrates but could be increased to 714 U/ml when combined with yeast and inoculum. The high yield of enzyme from the low cost areca husks as source for biomass was suggested to be beneficial for production of ethanol from various

Table 3.10 Properties of biochar obtained from areca husks compared to activated carbon and other sources of biochar (Mishra 2021)

Parameter	Areca husk	Activated carbon	Samanea saman biochar	Coal char	Palm shell hydro char
Moisture, %	4.27 ± 1.3	21.08 ± 0.8	5.14 ± 0.4	–	–
Volatile matter, %	22.83 ± 1.2	16.69 ± 1.2	31.14 ± 0.8	–	–
Ash content, %	10.79 ± 0.60	7.86 ± 0.6	13.18 ± 0.4	–	–
Fixed carbon, %	62.12 ± 1.4	54.37 ± 0.6	47.54 ± 0.8	–	–
C, %	73.14 ± 0.2	58.55 ± 0.2	62.66	55.38	63.77
H, %	2.50 ± 0.14	0.90 ± 0.1	2.06	5.86	4.40
O, %	22.82 ± 0.16	12.96 ± 0.14	31.83	34.07	23.33
N, %	1.54 ± 0.01	0.59 ± 0.01	3.45	2.48	0.52
S, %	–	–	–	2.21	1.02
O/C	0.23	0.11	0.38	0.46	0.27
H/C	1.31	0.83	0.40	1.27	0.83
Heating value, MJ/kg	29.11 ± 1.2	24.77 ± 1.1	23.14 ± 1.2	22.54	26.80
BET surface area m^2/g	5.49 ± 0.6	873.21 ± 0.8	8.20 ± 0.6	–	12.26
Bulk density, kg/m^3	280 ± 1.2	221.0 ± 1.8	478 ± 1.6	–	–
Acidity	7.79 ± 1.1	4.26 ± 0.2	7.60 ± 0.8	–	–
Zeta potential, mV	−31.04 ± 1.6	−29.05 ± 0.3	–	–	–

Reproduced with permission from Elsevier

sources of biomass (Sreena 2020). Areca husks were also found to be suitable for the production of cellulases using *Trichodema viride* through solid-state fermentation. Optimum conditions for enzyme production were pH of 5.75 and culture time of 5 days when an enzyme activity of 500 mg/ml/min was obtained (Bhagyamma 2011). In a similar study, it was found that arecanut husks could act as substrate for production of manganese peroxidase by two fungus *Phanerochaete chrysosporium* and *Phanerochaete sp.* Peroxidase with activity of 52.6 IU/g and 44.08 IU/g were obtained from the two microorganisms with optimum pH being 6 and 5, respectively. It was suggested that the enzymes could be produced from areca husks under ambient conditions and without the need for additional ingredients. A yield of up to 13% with 2.7 times purification was possible, and the specific activity increased up to 253 IU/g (Rajan 2020). Treating areca husks with *Pseudomonas fluorescens* at 36 °C for 3 days at pH 6 was able to generate cellulase with yield of 1.31 U/ml without adding external carbon source (Shashidhara 2013).

Table 3.11 Comparison of the physiochemical properties of pyrolysis oil obtained from areca husks (Mishra 2024)

Analysis	Thermal	ANH + ZSM-5	Diesel	Gasoline
Color	Black and brown	Black and brown	Light yellow	–
Yield, %	38.62 ± 1.61	49.63 ± 2.11	–	–
C (%)	59.36 ± 1.02	73.72 ± 1.41	87.40	–
H(%)	7.52 ± 0.86	9.08 ± 1.11	12.10	–
O(%)	30.64 ± 1.50	15.44 ± 2.41	–	–
N(%)	1.89 ± 0.21	1.56 ± 0.82	392	–
S(%)	0.59 ± 0.06	0.5 ± 0.02	1.39	–
HHV (MJ/kg)	26.45 ± 0.12	38.16 ± 0.16	45.50	47.30
Viscosity at 30 °C Cst	79.96 ± 0.42	36.56 ± 0.21	2–4.5	0.12
Density, kg/m^3	980 ± 4.49	812 ± 2.89	828	838
Moisture, %	2.10 ± 0.15	3.2 ± 0.15	0	<0.10
Acidity	6.89 ± 0.07	7.2 ± 0.04	5.5–8.0	–
Carbon residue, %	0.37 ± 0.02	0.43 ± 0.04	–	–

Reproduced with permission from the American Chemical Society through open access publishing

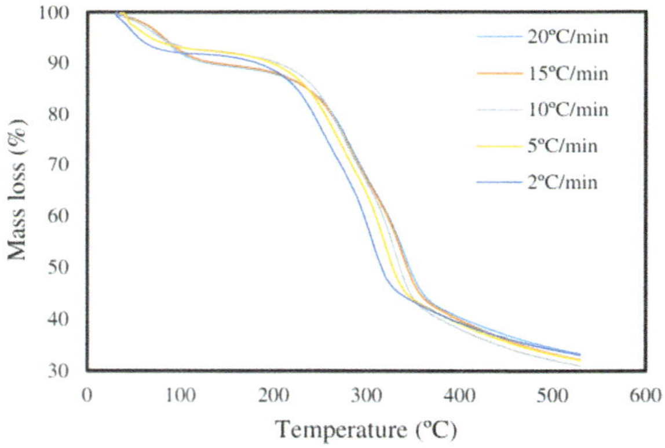

Fig. 3.13 TGA curve of areca husks subject to different heating rates (Gokul 2019). Reproduced with permission from Taylor and Francis

The possibility of generating ethanol from areca husks was studied using two types of fungi (AT—*Aspergillus terreus*, PC—*Phanerochaete chrysosporium* NCIM 1197) through separate hydrolysis and fermentation steps. Up to 58 (mg/g) of reducing sugar yield was obtained depending on the pH, incubation time, and other conditions. A combination of both fungi was found to provide better reducing sugar yield. The sugars obtained were fermented into ethanol using various bacteria and yeasts. Upto 10.7 g/L of ethanol yield could be obtained depending on the conditions (Table 3.13) (Aisabor 2016). Pretreatment which controls the extent of conversion of the polysaccharides into reducing sugars, was suggested to be critical during the production of

Table 3.12 Physiochemical properties of biochar obtained from raw and torrefied arecanut husks (Gogoi 2017) after heating at three different temperatures

Temperature, °C	Raw husks			Torrefied areca husks		
	300	400	500	300	400	500
Moisture content, %	6.5	5.8	5.4	4.8	4.3	3.7
Volatile matter, %	25	22.5	19.2	23.3	20.2	18.6
Ash, %	8.2	9.6	10.3	8.8	9.7	10.5
Fixed carbon	60	62.1	65.1	63.1	65.8	67.2
Higher heating value (MJ/kg)	–	–	28.49	–	–	29.76
C, %	67.22	71.75	73.14	69.53	73.55	74.77
H,%	4.23	4.02	3.37	4.08	3.89	3.23
N, %	3.98	3.59	3.60	2.75	2.80	2.62
O, %	24.57	20.64	19.89	23.64	19.76	19.38
H/C, %	0.76	0.67	0.55	0.70	0.63	0.52
O/C, %	0.27	0.22	0.20	0.25	0.20	0.19
pH	9.23	9.86	10.25	9.33	9.57	10.37
EC (dS/m)	2.9	3.1	3.5	3.1	3.30	3.80
R50 index	–	–	0.62	–	–	0.67
CSP%	–	–	25.26	–	–	32.98

lignocelluosic ethanol. The pretreatment affects the yield, cost, and technical requirements for conversion of the sugars into ethanol. The influence of alkali and acidic pretreatments on the conversion of areca husks into reducing sugars was studied by Naveenkumar et al. Areca husks were dried and powdered into three different particle sizes and subject to either alkaline or acidic hydrolysis. For the acidic hydrolysis, the powdered husks were treated with 1N sulfuric acid and heated at boil for 6 h, whereas the husk powder was treated with 0.25 M sodium hydroxide solution for 1 h. Extent of hydrolysis of the husks and the amount of reducing and non-reducing sugars formed (Table 3.14) showed that acid hydrolysis was more effective than alkaline hydrolysis. It was suggested that the extent of hydrolysis obtained would be suitable for commercial-scale production of ethanol from the husks (Naveen Kumar 2012).

Treating areca nut husks with different chemicals, such as acid (sulfuric acid, sodium hydroxide, hydrogen peroxide, and ethanol in sulfuric acid) at different temperatures resulted in the production of methane (Fig. 3.14). Alkaline pretreatment at 90 °C produced maximum biogas yield of 683 mL/g, about 2.3 times higher than that obtained using untreated husks. Among the biogas produced, methane constitutes about 71–75% depending on the parameters used during gasification (Vannarath 2022) (Fig. 3.15).

Table 3.13 Yield of ethanol (g/L) from arecanut husks treated with *Aspergillus terreus*, PC—*Phanerochaete chrysosporium*, and subject to fermentation by different fungi and yeasts at different inoculum concentrations (Aisabor 2016)

Inoculum concentration, %	*S. cerevisiae*	*S. uvarum*	*S. pombe*	*P. stiptis*	*C. shehatae*	*Z. mobilis*
1	8.8 ± 0.66	4.4 ± 0.12	4.8 ± 0.07	9.9 ± 0.85	8.3 ± 0.61	8.3 ± 0.41
2	9.4 ± 0.81	4.8 ± 0.09	5.2 ± 0.09	10.4 ± 0.94	8.8 ± 0.66	8.9 ± 0.31
3	9.8 ± 0.92	5.3 ± 0.23	5.8 ± 0.06	10.9 ± 0.45	9.7 ± 0.25	9.8 ± 0.42
4	8.7 ± 0.25	4.0 ± 0.14	4.6 ± 0.08	9.7 ± 0.23	8.4 ± 0.36	9.0 ± 0.51
5	8.5 ± 0.69	3.8 ± 0.08	4.4 ± 0.03	9.5 ± 1.0	8.0 ± 0.29	8.8 ± 0.50

Reproduced with permission through open access publishing

Table 3.14 Hydrolyzed products obtained (mg/g) from areca husks that were powdered to three different sizes and subject to acid and alkali hydrolysis (Naveen Kumar 2012)

Chemical constituent	Acid hydrolysis			Alkali hydrolysis		
	0.28 mm	0.43 mm	0.64 mm	0.28 mm	0.43 mm	0.64 mm
Reducing sugars	4.12 ± 0.02	3.18 ± 0.02	3.08 ± 0.01	1.6 ± 0.10	1.18 ± 0.02	0.86 ± 0.02
Non-reducing sugars	0.35 ± 0.03	0.37 ± 0.01	0.24 ± 0.01	1.02 ± 0.03	0.74 ± 0.03	0.72 ± 0.02
Protein content	3.35 ± 0.02	5.80 ± 0.10	5.95 ± 0.06	7.40 ± 0.15	6.95 ± 0.02	7.0 ± 0.12
Total carbohydrates	2.61 ± 0.03	2.82 ± 0.01	2.79 ± 0.01	1.83 ± 0.15	2.43 ± 0.02	1.86 ± 0.12

Reproduced with permission through open access publication

3.3.4 Xylose Production

The effect of treating areca husks with lime or dilute acid on the physical and chemical properties was investigated, and the treated husks were used for the production of xylose. Lime treatment was done by combining 1.25 g lime and 5 g of husks and incubating at 30 °C for seven days. Similarly, the husks were also treated (separately) with sulfuric acid and hydrolyzed at 121 °C for 30 min. FTIR spectrum (Fig. 3.16) and Raman spectroscopy analysis of the untreated and treated husks showed that cellulose and lignin were removed, and the fibers had high concentrations of hemicelluloses (29%). The lime and acid treatments had caused a substantial decrease in crystallinity, which could be preferable for production of the enzymes. When used for production of xylose, up to 85% yield (Fig. 3.17) was possible with acid treatment providing considerably higher yields than lime treatment (Vardhan 2020). Further, acid-treated husks provided 90% reducing sugar at an enzyme loading of 15 IU/g with average sugar yield of 104% depending on the enzyme concentration

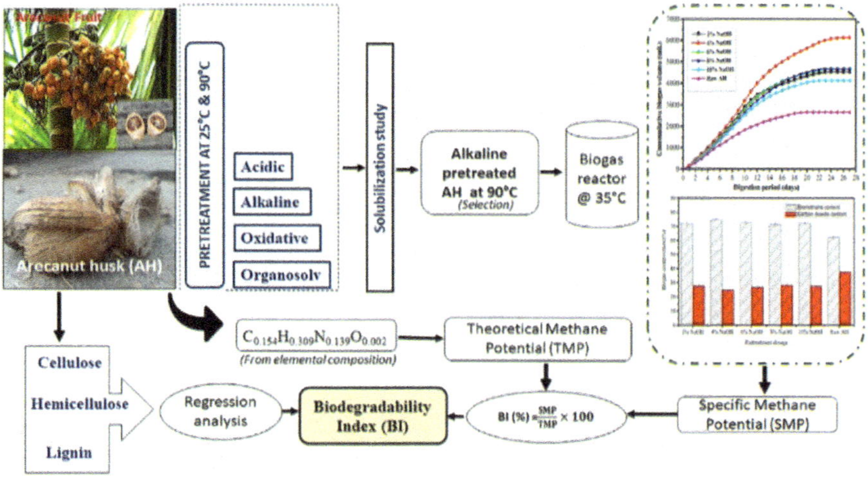

Fig. 3.14 Process used to convert arecanut husks into biogas (methane) (Vannarath 2022). Reproduced with permission from Elsevier

used (Vardhan 2024). Although areca nut husk (ANH) holds promise as a viable biomass source for xylose production, the recalcitrant nature of ANH makes saccharification more complex. To address this, lime and acid pretreatments were carried out to enhance the susceptibility of biomass to saccharification. Saccharification was conducted under the following conditions: 2% (wV^{-1}) substrate loading, 100 rpm agitation, and 30 °C hydrolysis temperature for 12 h hydrolysis time at pH 4.5–5.0. However, parameters like xylanase enzyme loading were varied to enhance the saccharification of the ANH. Results demonstrated that acid-treated husk (ATH), lime-treated husk (LTH), and raw husk (RH) achieved the highest yield (gg^{-1}) of reducing sugar, approximately 90, 83, and 15%, respectively, at an enzyme loading of 15.0 IUg^{-1}. Various analytical techniques, including Raman spectroscopy, Fourier transform infrared spectroscopy (FTIR), zeta potential, thermogravimetric analysis (TGA), X-ray diffraction (XRD), and field emission scanning electron microscopy (FESEM), were used to examine structural changes in the native, pretreated, and saccharified residues of ANH. The analysis revealed that a significant amount of partial crystalline and amorphous cellulose in the ANH biomass was hydrolyzed during the saccharification process. However, saccharification also led to the removal of amorphous substances, disruption of the crystalline structure, and conversion of crystalline regions into amorphous domains (Table 3.15).

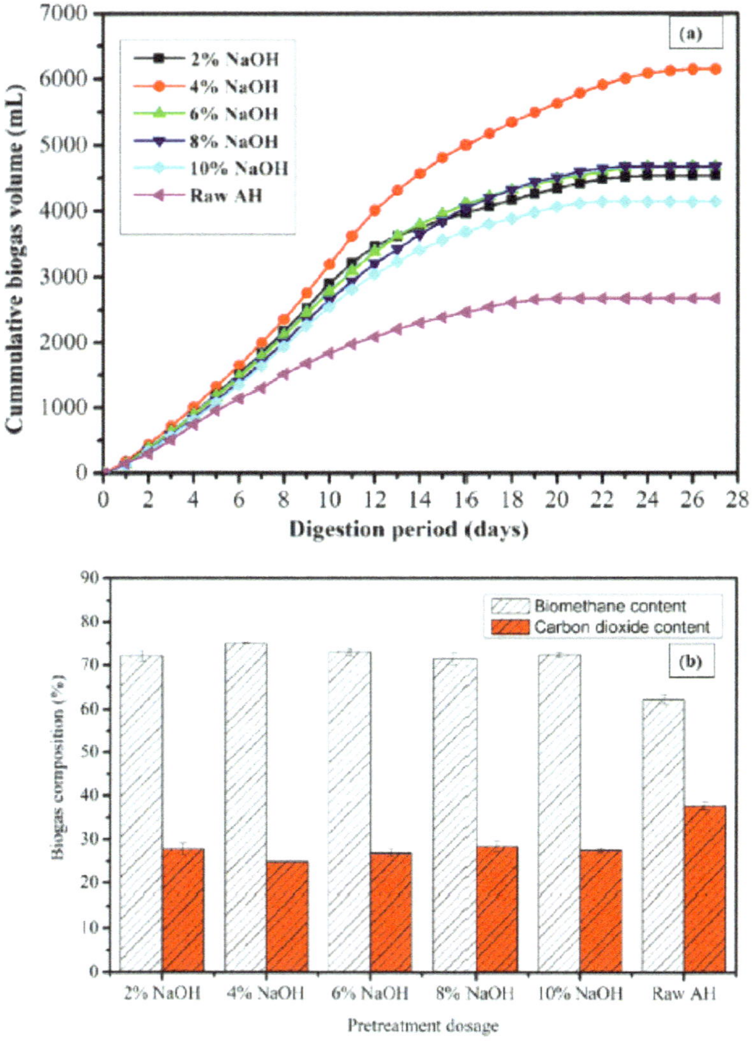

Fig. 3.15 **a** Cumulative biogas yield over the 27 days of the digestion period. **b** Biogas compositions for different pretreated AH (Vannarath 2022). Reproduced with permission from Elsevier

3.4 Natural Cellulose Fibers from Areca Husks

Areca husks are lignocellulosic with typical composition being 30–35% cellulose and 20–30% lignin which makes them an ideal source for natural cellulose fibers. However, the presence of high amounts of lignin makes it difficult to extract good quality fibers and develop high value fibrous applications for the husks. Since the

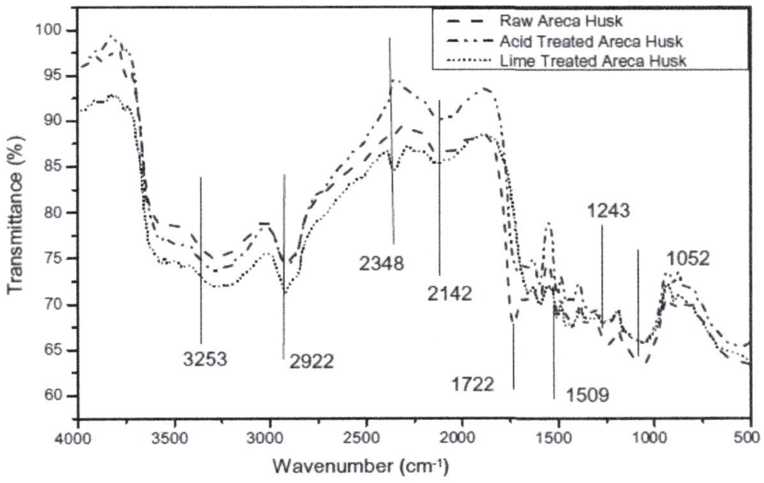

Fig. 3.16 FTIR spectra of raw areca husk, acid-treated areca husk, and lime-treated areca husk (Vardhan 2022). Reproduced with permission from Taylor and Francis

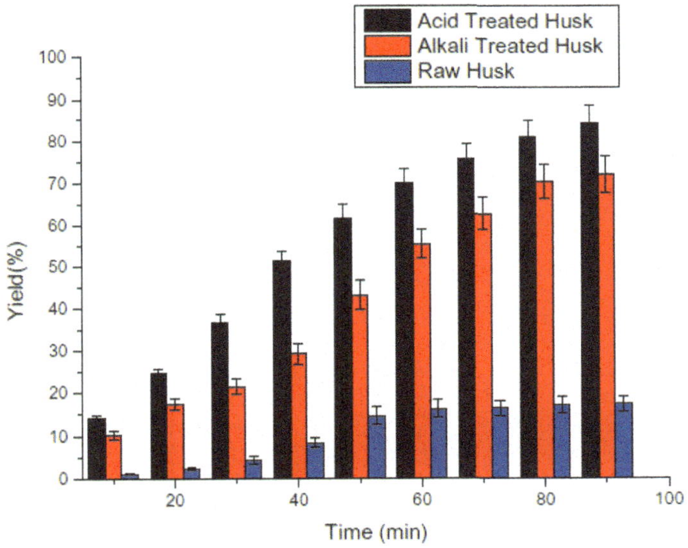

Fig. 3.17 Saccharification profile of raw and pretreated (acid and alkali) areca nut husk (Vardhan 2022). Reproduced with permission from Taylor and Francis

Table 3.15 Properties of the pure and areca nut husk nanoadditive containing biodiesel (Saha 2023)

Properties	Unit	Test method	Diesel	D10EPO	D10EPO-5 ppm	D10EPO-10 ppm	D10EPO-15 ppm
Density at 25 °C	Kg/m^3	ASTM D287	820	801	806	811	816
Calorific value	MJ/kg	ASTm D4309	44.3	42.3	42.5	42.6	42.74
Kinematic vis	cSt	ASTM D445	2.3	2.92	2.94	2.96	2.97
Flash point	°C	ASTM D93-58 T	53	50	52	54	57
Fire point	°C	ASTM D93-58 T	56	55	57	60	62
Cloud point	°C	ASTM D2500	−6	−4	−4	−4	−4
Pour point	°C	ASTM D97	−15	−12	−12	−12	−12
Cetane index	–	ASTM D61305	55	56	56	56	56

Reproduced with permission from Elsevier

length of the fibers in the husks is also low, appropriate means have to be developed to achieve optimum delignification and fiber quality. Since chemical delignification is harsher and it is difficult to control lignin removal, the possibility of treating the husks with enzymes and obtain natural cellulose fibers was studied. Husks were treated with various microorganisms based on their ligninolytic profile and two microorganisms *P. chrysosporium* and *Phanerochaete* sp were chosen and conditions were optimized to obtain biosoftened fibers. Tensile properties of the fibers had increased after treating with the microorganisms with strength increasing from 93 upto 125 MPa and elongation from 3.7 to 5%. Substantial reduction in color and increase in softness of the fibers was also observed indicating the suitability of the fibers for high value applications (Rajan 2005). Fibers obtained from areca husks are considerably coarse, short and have relatively low tensile strength and elongation which makes them unsuitable for most textile applications. To overcome these limitations, areca husks were subject to a process called biosoftening by treating with fungi. Dry areca husk were incubated with *P. chrysosporium* and *Phanerochaete sp* at adifferent pHs for up to 10 days in 0.2 M sodium acetate and phosphate buffer (Rajan 2005). Tensile strength of the fibers increased from 93 up to 125 MPa and elongation from 3.7 from 5% which were considered suitable for textile applications. In another study, areca husks were subject to a series of mechanical and chemical treatments to obtain natural cellulose fibers with properties suitable for high value applications. Mechanical treatments included pressing and busting of the husks after which they were soaked in 2.5 KOH solution for 24 h at a husk to solution ratio of 1:20. Treated husks were washed several times and fibers extracted were collected and studied for their structure and properties. Alkali treatment assisted in separation of the fibers without affecting the strength significantly. A high yield of up to 57% with maximum strength of the fibers being 36.9 N suggested suitability of the approach for practical fiber extraction (Table 3.16) (Deshmukh 2019).

In addition to lower mechanical properties, areca fibers suffer from high moisture absorption and susceptibility to water. Treating husks with alkali and later esterification with acrylic acid were found to substantially decrease the moisture sorption from about 65 to 30% and also increase compatibility with the polymer matrix when used as reinforcement for composites (Sampathkumar 2012). Fibers were obtained from areca nut husks of different growing ages by treating them with water or alkali (Lazim 2014a). However, treating fibers extracted from unripe, ripe, or matured areca husks with 5% NaOH for 30 min resulted in reduction in tensile strength and modulus but increase in elongation at break (Table 3.17). Further, alkali treatment substantially increased the debonding force and interfacial shear strength (Lazim 2014b). Although the process of extraction of fibers was not clearly described, a detailed chemical and physical characterization of natural cellulose fibers from areca husks was done by Binoj, et al. Fibers obtained had length between 10 and 60 mm and diameter of 396–479 μm considerably shorter and coarser than coir or common lignocellulosic fibers. Interestingly, the density of the fibers was reported to be between 0.7 and 0.8 g/cm^3 substantially lower than most lignocellulosic fibers. Depending on the conditions during testing, tensile strength of the fibers varied from 148 to 323 MPa, elongation from 10 to 13%, and modulus from 1.1 to 3.2 GPa, considerably lower

Table 3.16 Conditions used and properties of fibers obtained from areca nut husks (Deshmukh 2019)

Mechanical treatment	Chemical treatment	Fibre recovery, %	Detachment force, single fiber N	Detachment force, fiber bundle N	Fibre strength, N
Pressing	Soaking in water for 24 h	57.34	8.03	15.48	34.19
	Soaking in 2.5% KOH for 24 h	52.77	5.12	13.27	36.85
Busting	Soaking in water for 24 h	49.45	10.58	17.32	34.10
	Soaking in 2.5% KOH for 24 h	44.14	7.29	17.34	36.16
Untreated	Soaking in water for 24 h	47.50	10.63	24.19	32.47
	Soaking in 2.5% KOH for 24 h	43.35	7.61	23.23	36.13
Mean of chemical treatments	Soaking in water for 24 h	51.43	9.74	18.99	33.58
	Soaking in 2.5% KOH for 24 h	46.75	6.67	17.94	36.38

compared to the common fibers. The fibers were reported to have a crystallinity of 55% and crystal size of 7.9 nm (Binoj 2016). However, the physical properties of fibers obtained from areca husks depend not only on the source but also on the type of chemical or physical treatment used for extraction (Table 3.18) (Loganathan 2020). For instance, substantial changes to the chemical, physical, and morphological features were observed when areca fibers were treated with various concentrations of alkali (Hu 2021) (Fig. 3.18 and Table 3.19).

Similar to the extraction of fibers, the possibility of converting the husks into pulp for production of rayon (regenerated cellulose) fibers was explored. Three (water pre-hydrolysis, soda cooking, and chlorine-free bleaching) methods were used to prepare the pulp. Soda cooking provided pulp with a kappa number between 16 and 18.4 and a yield between 29 and 42%. With 89% ISO brightness, α-cellulose content of 96%, and viscosity of 7.6 cP, the pulp was considered suitable for production of rayon (Fadli 2023).

Natural cellulose fibers obtained from areca husks after treating with 6% NaOH solution at room temperature (25–30 °C) for 24 h were blended with fibers obtained from water hyacinth. Fibers obtained were pulverized and used to develop biodegradable plates for food consumption with tapiaco starch, guar gum as the binder,

Table 3.17 Properties of natural cellulose fibers extracted from unripe, ripe, and matured areca husks before and after treating with alkali (Lazim 2014)

Areca fibers	Tensile strength, MPa	Young's modulus, MPa	Elongation at break, %
Untreated			
Unripe	124 ± 50	1286 ± 710	22.3 ± 6.5
Ripe	166 ± 55	1381 ± 552	23.2 ± 5.0
Matured	129 ± 44	2569 ± 637	23.1 ± 6.7
Treated with 5% alkali			
Unripe	33 ± 9	375 ± 86	46.4 ± 12.4
Ripe	45 ± 10	616 ± 136	32.5 ± 6.3
Matured	31 ± 6	364 ± 77	37.3 ± 8.5

Reproduced with permission through open access publishing

and glycerol as the plasticizer. Various ratios of the areca and hyacinth fibers were prepared and compression molded into plates at 110 °C for about 7 min (Rajbongshi 2025). The plates were coated with beeswax to increase hydrophobicity and performance properties. Some of the properties of the plates obtained are given in Table 3.20, and images of the plates are shown in Fig. 3.21. A 50/50 ratio of areca and hyacinth fibers was found to provide optimum properties with tensile strength of 3.99 and 5.01 MPa without and with coating. Plates without the coating were able to biodegrade within 6 months in both sand and soil.

3.5 Nanocellulose from Areca Husks

In addition to the extraction of natural cellulose fibers from the areca husks, the possibility of using the husk to extract nanocellulose was studied by Chandra et al. The procedure used to obtain the nanocellulose from the areca husks is shown in Fig. 3.19. Nanocellulose obtained had a α-cellulose content of 85%, lignin content of 2.3% compared to 34% and 32% in the raw husks, respectively. Morphological analysis showed that most of the nanocellulose was in nanofibrillar form with average fiber diameter of 3–5 nm and aspect ratio between 120 and 150. Nanocellulose showed substantially high thermal stability and the crystallinity had increased to 73% from 37% and the crystalline domain size increased from 2.8 to 4.3 nm (Chandra 2016). Properties of the nanocellulose obtained were considered suitable for industrial applications. In another study, areca nut husk was used to extract cellulose nanocrystals with diameter of 19 nm, length of 195 nm, and zeta potential of −15.3 mV. Addition of the nanocrystals into poly(vinyl alcohol) or chitosan improved the tensile strength by about 21%, and the films showed good antimicrobial activity and were hence considered suitable for food packaging applications (Perumal 2022). In a unique approach, nanofibrillated cellulose fibers obtained from areca husks were self-assembled into giant vesicles by treating with oleylamine (Soman 2018). The

Table 3.18 Properties of fibers obtained from areca husks using different extraction techniques in comparison to coconut and palm fibers (Loganathan 2020)

Fibres	Length, mm			Diameter, μm	Ratio (Length/ diameter)	Density, g/ cm^3	Crystallinity index, CI %
	Short	Medium	Long				–
Arecahusk fiber	18–29	30–38	39–46	28–89	586	1.05–1.25	–
Areca husk fiber	–	10–60	–	396–476	–	0.75 ± 0.03	–
Single betel nut fiber	–		–	32	–		–
Areca frond fibre	–	200–700	–	–		1.14–1.20	
Untreated areca leaf stalk fiber	–	–	–	285–330	–	1.09 ± 0.02	–
5% alkali treated areca leaf stalk fibres	–	–	–	260–315	–	1.17 ± 0.035	–
10% alkali treated areca leaf stalk fibres	–	–	–	225–290	–	1.28 ± 0.029	–
15% alkali treated areca leaf stalk fibres	–	–	–	210–262	–	1.34 ± 0.027	
Untreated raw husk	–	–	–	–	–	1.06	–
Untreated ripe husk	–	–	–	–		1.16	–
Untreated matured husk	–	–	–	–	–	1.46	–
Areca	10–60	–	–	396–476	80	0.7–0.8	55.5
Coconut	20–150	–	–	100–460	303	1.15–1.46	–
Palm	–	–	–	150–500	–	0.70–1.55	34.1

Reproduced with permission from Springer Nature

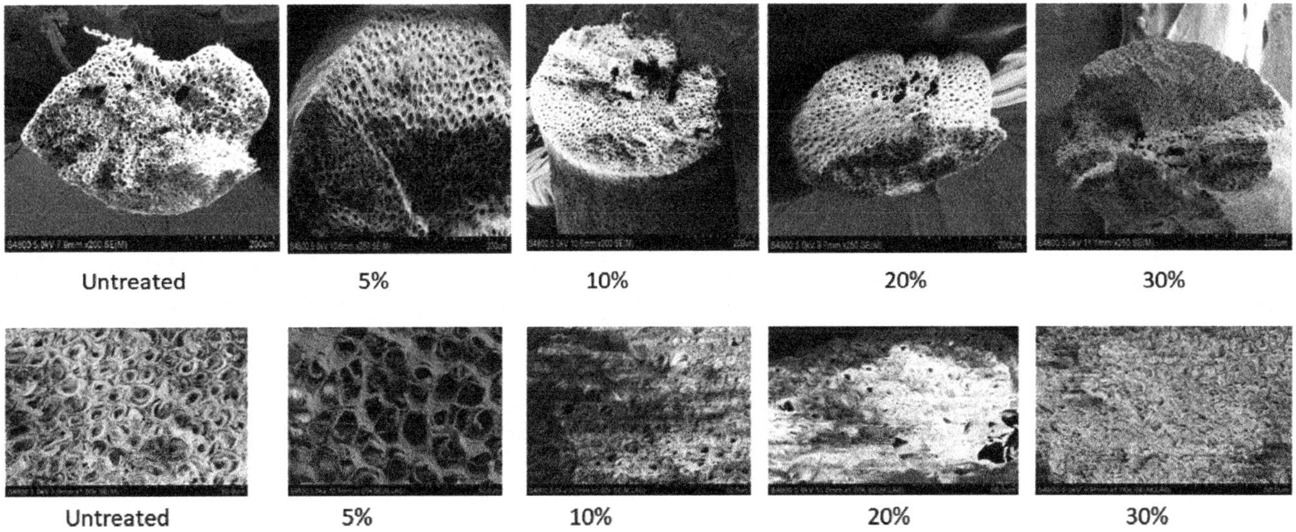

Fig. 3.18 Scanning electron microscope images of untreated fibers and those treated with four different concentrations of alkali for 24 h. Top row is images taken at 200x and bottom row images taken at 1000x (Hu 2021)

Table 3.19 Properties of natural cellulose fibers from areca shells obtained at different extraction conditions in comparison to coir, cotton, and jute fibers (Hu 2021)

Treatment		Tensile strength (MPa)	Elongation at break (%)	Density (Tex)
Time, hours	Alkali concentration, %			
24	Raw fiber (Untreated)	23.82 ± 12.6	34.3 ± 12.25	109.0 ± 0.56
	5	52.29 ± 9.8	44.86 ± 8.45	94.8 ± 0.63
	10	58.70 ± 8.9	39.73 ± 9.92	71.4 ± 0.59
	20	35.96 ± 8.8	28.77 ± 6.26	63.4 ± 0.74
	30	34.67 ± 9.3	36.23 ± 9.66	60.3 ± 1.64
48	5	26.64 ± 8.2	41.10 ± 18.47	84.8 ± 0.65
	10	27.45 ± 9.3	46.20 ± 12.88	56.6 ± 0.78
	20	37.90 ± 9.4	31.70 ± 5.37	50.2 ± 0.48
	30	59.03 ± 8.6	36.62 ± 9.02	49.5 ± 0.59
Coir	Mechanical extraction	204–234	20–40	120–220
Cotton	–	360–570	6–9	8–9
Jute	–	393–500	2–3	1–5

nanofibers were obtained from the areca husks using a TEMPO-mediated oxidation process. Later, the nanofibers were modified with oleylamine at two different concentrations. After treatment, the cellulose papers showed an amidation % between 16 and 69 with contact angle increasing from 44 up to 110°. During the amidation process, the cellulose nanofibers were self-assembled into nanovesicles between 2 and 5 μm diameter and thickness between 300 and 600 nm. The self-assembly was suggested to be due to the folding of the fibrils through non-polar interactions of the cis-unsaturated long aliphatic chains and stabilized by the intermolecular hydrogen bonding. The vesicles formed were suggested to be suitable for storage of drug and cosmetic materials and for developing biosensors (Soman 2018).

3.6 Miscellaneous Applications

In a different approach, areca husks were made into ash and were chemically modified with lithium to form a solid catalyst for biodiesel production from waste cooking oil. In this approach, areca husks were heated at 750 °C for 3 h, and the ash formed was combined with lithium nitrate with citric acid as a surfactant. This mixture was calcined at 650 °C for 3 h and purified and reactivated by treating at 650 °C for 1 h. The catalyst thus formed was used for production of biodiesel. FTIR and XRD analysis (Fig. 3.20) indicated that the catalyst was in the form of lithium silicates such as Li_2SiO_3 and Li_2SiO_5. When used for production of biodiesel, the catalyst showed high heterogeneous behavior and tolerance for low levels for free

Table 3.20 Properties of the biodegradable plates developed using areca husk and hyacinth fiber (Rajbongshi 2025)

Areca/hyacinth fiber ratio	Thickness (mm)	Grammage (g m^{-2})	Bulk density (g cm^{-3})	Moisture content (%)	Water absorptiveness (g m^{-2})	Absorbency (s)
100/0	1.66 ± 0.19^{a}	994.33 ± 38.00^{a}	0.127 ± 0.019^{a}	9.27 ± 1.99^{a}	248.00 ± 50.46^{a}	5507.00 ± 202.07^{e}
80/20	1.68 ± 0.04^{a}	1090.00 ± 13.45^{b}	0.150 ± 0.019^{b}	10.62 ± 0.25^{b}	309.67 ± 55.53^{ab}	4632.00 ± 281.91^{d}
60/40	1.82 ± 0.05^{ab}	1220.23 ± 23.41^{c}	0.146 ± 0.006^{b}	11.18 ± 0.24^{bc}	388.90 ± 15.50^{bc}	592.33 ± 90.97^{c}
50/50	1.93 ± 0.06^{bc}	1315.67 ± 20.55^{d}	0.184 ± 0.006^{c}	10.27 ± 0.26^{b}	481.33 ± 61.04^{c}	408.00 ± 17.09^{bc}
40/60	1.95 ± 0.12^{bc}	1276.00 ± 28.16^{cd}	0.193 ± 0.004^{c}	13.59 ± 0.11^{d}	811.67 ± 45.65^{d}	191.00 ± 52.37^{ab}
20/80	1.98 ± 0.04^{bc}	1401.67 ± 80.41^{e}	0.196 ± 0.003^{c}	12.85 ± 0.24^{d}	1380.67 ± 117.37^{e}	181.67 ± 14.05^{ab}
0/100	2.05 ± 0.06^{d}	1483.33 ± 70.78^{f}	0.204 ± 0.005^{c}	12.47 ± 0.35^{cd}	1934.67 ± 113.57^{e}	129.33 ± 15.57^{a}

Reproduced with permission through open access publishing. Data points with different alphabets indicate statistically significant difference

Fig. 3.19 Process of preparing the cellulose nanofibrils from areca nut husks (Chandra 2016). Reproduced with permission from Elsevier

fatty acid (FFA) and water content in the waste cooking oil. Only 1% catalyst was sufficient for transesterification, and the catalyst also showed recyclability for up to three cycles (Vinu 2020). In a related application, nanoadditives obtained from areca husks were combined with plastic grocery bag waste, and the effects on emission, combustion, and engine performance were monitored. Depending on the concentration (5–15 ppm), of the nanoadditive, the peak heat release rate, in-cylinder pressure were found to decrease. About 9.1% decrease in fuel consumption and reduction in smoke emissions were observed specifically for the 10% additive (Saha 2023). Some of the properties of the arecahusk nanoadditive containing biodiesel are given in Table 3.15. In a different approach, activated carbon obtained from areca husks was made into reduced graphene oxide and used as a nanoadditive to increase the catalytic process and combustion rate in internal combustion diesel engines. Adding 25 ppm of areca husk graphene oxide provided better combustion and emission parameters and lowered HC emissions by 43%, CO emissions by 33%, NOx by 2.4% and smoke emissions by 15% (Muralidharan 2022). Another study has also confirmed the beneficial effects of adding areca nanoadditive (surface area of 3 m^2/g and pore size of 4.9 nm). The additive improved the energy and exergy efficiency up to about 25%. At 15 ppm, the additive was found to provide the best efficiency and highest sustainability index (Saha 2024). Similarly, Yadav et al. used areca husk to synthesize a solid catalyst with high acid density through low-temperature hydrothermal carbonization. The catalyst had high levels of sulfonic active sites and high acid density and was used for the production of biodiesel from oleic acid. A highly porous structure (diameter of 2.21 nm, pore volume of 0.023 cc/g, and surface area of 19.561 m^2/g) and the presence of sulfonic groups provided high activity for the catalyst with 85% transformation of oleic acid to methyl oleate even after 4 cycles (Yadav 2022).

Biochar produced from areca husks has been considered for soil quality management and promoting sustainable agriculture. In one such attempt, areca husks were carbonized between 250 and 400 °C for 30–90 min. The biochar (particle size of 1191 nm), having highly oxygenated functional groups and porous structure, formed high soil organic matter yield index and lowest hydrogen/carbon and oxygen/carbon ratios. When applied to soil, the biochar increased plant height and water content of vetiver (Gujre 2022).

In a similar approach, areca husks were made into activated carbon and later combined with TiO_2 to develop electrodes for the detection of the common pesticide dichlorophen. Activated carbon obtained was found to have an average size of 100 nm with amorphous and irregular structure. Combining the carbon with TiO_2 leads to a sixfold increase in peak response. An irreversible electrochemical response governed by a diffusion-controlled process was observed. The sensor showed good sensitivity with DL and QL values of 1.01×10^8 M and 3.36×10^8 M, and was considered suitable for detection of pesticides in soil and other environmental samples (Pai 2023).

In a unique study, the potential of using areca husks as a source for zinc silicate was explored (Anuar 2021). Initially, areca husks were heated between 700 and 900 °C, and the silica powder obtained was crushed to obtain particles of 45 μm in size. Zinc silicate was prepared from the area silica using the procedure shown in Fig. 3.21. Compositional analysis showed that the silicate obtained was mainly

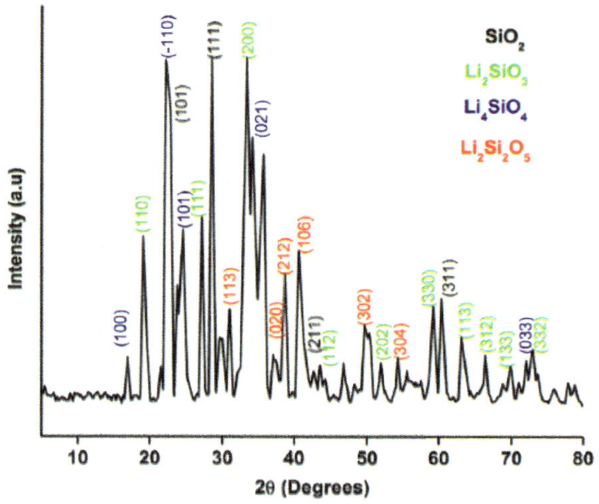

Fig. 3.20 XRD pattern of the lithium silicate derived from areca husk ash for potential use as catalyst for production of biodiesel from waste cooking oil (Vinu 2020). Reproduced with permission from Elsevier

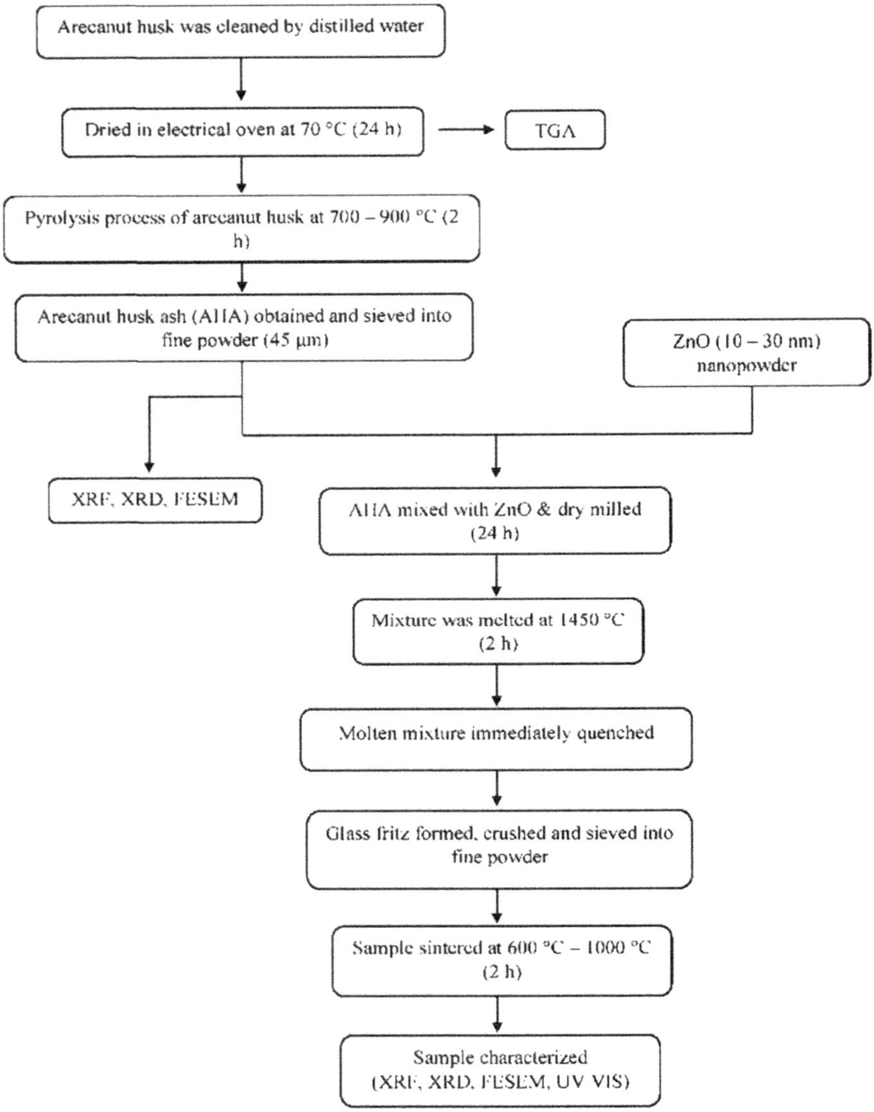

Fig. 3.21 Schematic of the steps used in the preparation of the areca nut ash and zinc silicate (Anuar 2021). Reproduced with permission through open access publication

composed of SiO_2 (45.4%) and P_2O_5 (14.16%) when prepared at 900 °C. The SiO_2 had purity between 29 to 45% with high crystallinity. The zinc silicate had grain size between 1011 and 3518 nm and an optical band gap between 2.410 and 2.697 eV. Areca husks were considered a green and low-cost source for the production of phosphorous materials (Anuar 2021).

In addition to enzymes and biofuels, areca husks have also been used to produce chemicals. In one such attempt, the feasibility of producing citric acid from areca husks using solid-state fermentation and *Aspergillus niger* as the microorganism was studied (Narayanamurthy 2008). In this study, areca husks containing 27% sugars and 37% moisture were sterilized and inoculated with 10% pore suspension and incubated at 30 °C, and the production of citric acid was measured under different conditions. Under the optimum pH of 5 and 50% moisture in the husks, a total citric acid yield of 119 g per kilogram of husk used, corresponding to a citric acid yield of 67% and sugar utilization of 85% was obtained.

3.7 Conclusions

Areca husks are major byproducts inevitably generated during areca processing. Although available in large quantities and low cost and with limited applications are present, presence of high lignin content and short and coarse fiber structure makes processing areca husks difficult. Nevertheless, studies have clearly demonstrated the benefits of using areca husks as sorbents for dyes, heavy metals, and other pollutants. Activated carbon prepared from areca husks has high electrical performance and is found suitable for energy storage applications. It has also been shown that areca husks can be used directly as fuel or converted into ethanol. Similarly, areca husks have been used as substrates for enzyme production and as a source for natural cellulose fibers. Areca husks have shown the potential for multiple applications and could replace the synthetic polymer-based materials. Using areca husks could also benefit the farmers and support green economy, sustainability, and other initiatives.

Chapter 4
Areca Husks and Fibers from Husks as Reinforcement for Composites

4.1 Introduction

Areca husks are renewable and sustainable sources and are generated as byproducts when processing the areca nuts. Hence, large quantities of the husks are available at relatively low cost. Areca husks are lignocellulosic and highly fibrous in nature, and hence suitable for use as reinforcement for composites similar to other natural fibers and biomass. Considerable research has been published on using areca husks in their native form and also after various physical and chemical treatments as reinforcement for composites. In addition to directly using the husks, areca husks have been physically and/or chemically treated to obtain fibers, and in turn, the fibers have been used as reinforcement for composites. Some of the treatments and/or conditions used for developing areca husk composites are shown in Table 4.1 (Makruf 2024; Loganathan 2020; Gokarneshan 2022). Composites are essentially composed of two major components, the matrix and reinforcement. Based on the type of matrix and reinforcement used and the properties desired, additives, compatibilizers, and other constituents may also be incorporated in the composites. Depending on the type of matrix, composites can be classified as either being completely biodegradable or partially biodegradable. Completely biodegradable composites are made using both reinforcement and matrix from biodegradable sources, whereas partially biodegradable composites are made with either the matrix or reinforcement from biodegradable sources. Typically, biocomposites are prepared using natural fibers and reinforcements and synthetic or biodegradable matrices. In the case of areca husks and fibers, both synthetic and biodegradable matrices have been used, leading to completely biodegradable and partially biodegradable areca husk composites. In this chapter, the various types of areca husk-based composites developed are classified based on the type of matrix used, and the structure, properties, and applications of the composites are discussed.

N. Reddy and P. Aramwit, *Areca catechu and its byproducts*,
https://doi.org/10.1007/978-3-032-22134-6_4

Table 4.1 Comparison of the different treatment conditions used to obtain areca husk fibers for composite applications and the properties of the composites obtained with different matrices (Makruf 2024). Reproduced with permission through open access publishing

Chemical treatment				Composition and properties				
Chemical	Concentration (%)	Duration, h	Temperature, °C	Fiber, %	Matrix	Additives	Tensile strength, MPa	Flexural strength, MPa
NaOH	20	1	–	40	Epoxy	2% Biochar	178	282
NaOH	10	36	30	40	Epoxy	2% Biochar	178	188
NaOH	6	24	26	20	Epoxy	–	28,24	52,11
NaOH	5	2	RT	30	PP	–	27	63
NaOH	5	1	80	65	Epoxy	–	25,9733	–
NaOH	5	0.5	RT	20	Polyester	20% Sponge gourd	22,06	40,1
NaOH	0.5	0.166	RT	05	Epoxy	–	15,6	–
KOH	2.5	24	RT	15	Epoxy	–	15,1	–
KOH	6	–	28–125	60	Epoxy	–	–	74,613
				03	PP	–	22	–
				07	Epoxy	–	–	–
				15	Epoxy	–	9,19	–

4.2 Areca Fiber Reinforcement for Epoxy Resin

Among the non-biodegradable synthetic polymer-based matrices, epoxy is one of the most common matrix used for preparing composites. Although non-biodegradable, the low processing temperature, good properties, and relatively low cost make epoxy a preferred binder. Hence, extensive reports are available on areca husk and areca husk fiber reinforced epoxy composites in individual and also as hybrids with other fibers (Ashok 2018). Due to the large availability, low cost, fibrous structure, and durability, several studies have been done on using areca nut husks with and without treatment as reinforcement for composites. In a study by Kamath and Bennehalli, areca husks were treated with 1% alkali to enhance compatibility with epoxy as the binder. Treated fibers were combined with epoxy and made into composites. Adding 50% of treated husk fibers provided optimum increase in properties with tensile strength of 14.85 MPa and flexural strength of 1.37 MPa, which were considered suitable for commercial applications (Kamath 2021). In another study, areca husks treated with 1% alkali were combined in different ratios (45–65%) in epoxy and made into composites. Addition of alkali-treated fibers provided better strength to the composites compared to adding untreated fibers, with tensile strength of 7.4 MPa and flexural strength of 4 MPa at 50% fiber reinforcement (Kamath 2021). In addition to mechanical properties, other studies have also shown that addition of areca husk fibers improves the thermal and electrical conductivity of epoxy-based composites with value of 0.22141 M^2 k/w, and 1.888×10^{10} Ω, respectively. Similarly, epoxy-areca husk composites also had good sound absorption coefficient of 0.7 at a frequency of 6400 Hz suitable for building applications (Srinivasan 2023). Untreated areca husks were made into a powder and added to polyurethane foams to improve acoustic properties. Addition of the fibers provided excellent acoustic properties with peak sound absorption value of 0.95 obtained at 450 Hz, considerably higher than those reported in literature (Santosh 2024).

Combining untreated and alkali-treated areca husk fibers having length between 40-50mm and diameter between 0.20 and 0.30 mm with epoxy resin resulted in composites with highest tensile strength of 7.4 MPa and flexural strength of 4 MPa when untreated fibers were used, compared to 54.9 and 6.8 MPa, respectively, for the alkali-treated fiber reinforced epoxy composites (Kamath 2021). It has been shown that epoxy provides composites with better properties (higher impact resistance and tensile properties) compared to melamine formaldehyde or urea formaldehyde because of their higher molecular weight (Srinivasa 2013). Epoxy reinforced with treated and untreated areca husk fibers showed good interaction when the fiber content was 20%. Composites had good thermal and structural resistance required for various applications (Sankarathil 2023). In addition to epoxy, a natural resin called dammar gum (melting point of about 140 °C) was combined together (up to 40% in epoxy) to develop composites using areca husk and banana fibers as reinforcement, with fiber content being constant at 10% for areca and between 10 and 30% for banana fibers. Tensile strength varied from 10 to 20 MPa, with composites containing 80% epoxy and 20% gum showing higher tensile strength and bending strength. Similarly, fibers containing 20% banana and 10% areca fibers provided highest bending resistance (Anand 2024). Composites intended for construction, interior, aerospace, and other applications were made using areca husk fibers as reinforcement and epoxy as the matrix. Higher storage (2.61 GPa) and loss modulus were obtained when 6% alkali treated fibers were used as the reinforcement (Mathapati 2022). In another study, areca husks were treated with 5% NaOH for 2 h and used as reinforcement for epoxy. A comparison of the composite properties was done using only mechanically extracted fibers with fiber volume varying from 10 to 30%. Some of the properties of the composites containing untreated or treated fibers are shown in Table 4.2 (Sankarathil 2023). Adding short (30 mm) areca fibers into epoxy resin (20%) was found to not only improve flexural strength and modulus but also impact strength and shore hardness, as reported in other literature as well (Palaniswamy 2024).

Incorporating 2 to 6% (w/w) of water-extracted short areca fibers into epoxy provided improved wear rate and tensile properties. The extent of improvement in properties could be estimated using finite element analysis (Choudhary 2021). In another study, alkali-treated and untreated fibers were used separately as reinforcement for epoxy composites, with 4% fiber containing composites providing highest tensile strength of 14.2 MPa compared to 10.1 MPa for the neat epoxy resin. Hardness of 55 was observed for 5% fiber containing composites compared to the unreinforced epoxy matrix (Borah 2018). Adding 40 to 60% short areca fibers into epoxy increased fracture load and fracture toughness, with higher fibre content providing improved properties (Manjunatha 2020). To obtain composites with properties suitable for various applications, areca husk fibers were extracted with alkali and treated with 4% benzoyl peroxide, and later combined with epoxy and made into composites. Chemical treatment of the fibers increased % crystallinity and thermal stability and improved the performance of the composites (Chandrappa 2022).

Table 4.2 Comparison of the properties of untreated (UT) and alkali-treated (AT) areca husk fiber reinforced epoxy composites with different extents of fiber contents in the composites (Sankarathil 2023). Reproduced with permission from Springer Nature

Fiber content, %		Tensile strength, MPa	Elongation, %	Tensile modulus, MPa	Flexural strength, MPa	Flexural modulus, MPa	Impact strength, KJ/m2	Hardness
10	UT	10.10	0.24	2913	31.76	2490	2.18	80
	AT	20.20	3.21	3356	42.35	2590	2.36	82
15	UT	12.31	0.3	3680	37.96	3243	2.48	80
	AT	24.62	3.13	4235	48.68	3363	2.98	82
20	UT	14.12	0.30	4903	41.12	4303	2.64	84
	AT	28.24	3.30	5012	52.11	4878	3.34	86
25	UT	10.87	0.34	3300	36.03	2963	3.82	81
	AT	21.74	3.24	3523	47.23	3012	4.12	83
30	UT	9.78	0.28	2840	34.96	2733	3.58	82
	AT	18.20	3.32	3012	43.60	2989	3.90	84

Instead of fibers, areca fibers in the form of mats having density of 300 gsm and tensile strength between 320 and 876 MPa were used in multiple layers as reinforcement for epoxy composites (Fig. 4.1). Tensile strength and modulus of the composites increased in proportion to the number of layers of areca mats whereas flexural and impact strength did not show a major increase when the number of layers was increased above one. However, water sorption increased from 7.7 to 9.9%, and the composites with higher amounts of areca fiber mats were also easily biodegradable. Morphological analysis showed good fiber interaction with the matrix, and minimum voids were observed in the composites. It was suggested that using areca fiber mats would be preferable for composites intended for automotive and other applications (Satyanarayana 2024). In a related study, areca husk fibers were made into mats and used as a reinforcement for epoxy resin. Storage modulus of 2.6 GPa at 28 °C and decrease in loss modulus were observed with increasing temperature (Aishwarya 2022). Similar to the effect of using different layers, chemical treatments were also found to influence the properties (specifically impact strength) of the areca fiber reinforced epoxy composites. Fibers treated with acrylic acid and composites containing 60% fiber loading had highest strength of 28 J compared to 10 J for composites containing untreated fibers (Dhanalakshmi 2015).

In another study, L12 epoxy resin and camcure-2389 hardener were used as matrix, and areca husk fibers treated with 6% NaOH for 24 h at 26 °C were used as reinforcement. Composites of 5 mm thickness and containing 15% fibers were obtained after curing at room temperature for 6 h. When subject to dynamic mechanical analysis (DMA) at different frequencies, flexural modulus of the composites was found

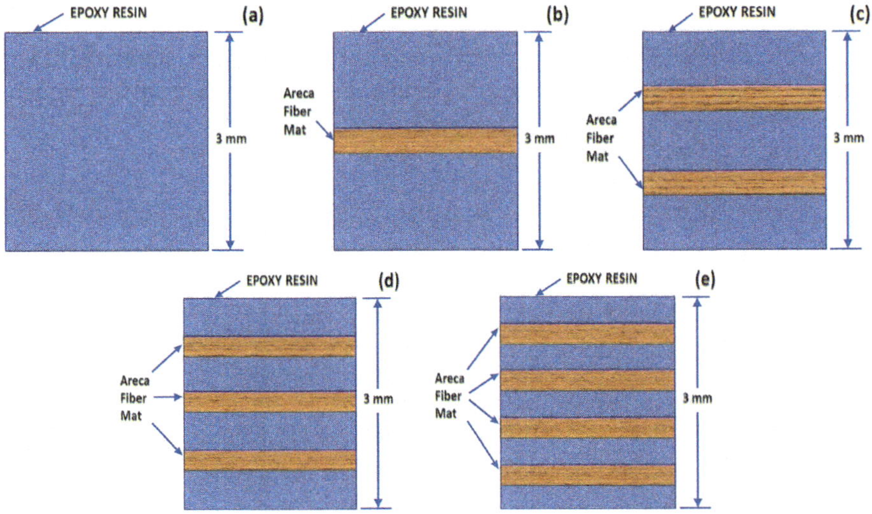

Fig. 4.1 Areca fiber mat composites: **a** neat polymer composite, **b** single-layer fiber mat, **c** double-layer fiber mat, **d** triple-layer fiber mat, and **e** four-layer fiber mat (Satyanarayana 2024). Reproduced with permission from Springer Nature

to vary between 2.7 and 3.4 GPa, with modulus increasing at higher frequencies (Muralidhar 2020). In a study by Kamath and Bennehalli, areca husks were treated with 1% alkali to enhance compatibility with epoxy as the binder. Incorporating about 50% of treated husk fibers provided optimum properties with tensile strength of 14.85 MPa and flexural strength of 1.37 MPa suitable for commercial applications (Kamath 2021). In another study, areca husks treated with 1% alkali were combined in different ratios (45–65%) of epoxy and made into composites. Addition of alkali-treated fibers provided better strength to the composites compared to adding untreated fibers with tensile strength of 7.4 MPa and flexural strength of 4 MPa at 50% fiber reinforcement (Kamath 2021). Other studies have also shown that addition of areca husk fibers improves the thermal and electrical conductivity of epoxy-based composites with value of 0.22141 M^2 k/w, and 1.888×10^{10} Ω with and without the fibers, respectively. Similarly, epoxy-areca husk composites also had good sound absorption coefficient of 0.7 at a frequency of 6400 Hz suitable for building applications (Srinivasan 2023). Composites intended for construction, interior, aerospace, and other applications were made using areca husk fibers as reinforcement and epoxy as the matrix. Higher storage (2.61 GPa) and loss modulus were obtained when 6% alkali treated fibers were used as the reinforcement (Mathapati 2022).

Several studies have also been done to treat areca husks with chemicals other than sodium hydroxide. In one such attempt, fibers were obtained from areca husks after treating with 10% potassium hydroxide at 30 °C for 36 h. These fibers were combined with epoxy resin and converted into composites. Alkali-treated fibers provided higher compressive and tensile stress and flexural strength of up to 60 N/mm^2. Properties of the composites varied considerably depending on the proportion of the fibers and the curing conditions adopted (Venkateshappa 2010, 2012). Instead of directly using areca fibers, the fibers were hand-spun into threads and later combined with epoxy and made into composites. Extensive analysis was done using dynamic mechanical analysis at different loading frequencies, and it was found that the storage modulus varied between 0.478 and 0.607 GPa, elastic modulus was between 2.7 and 3.4 GPa under dynamic testing conditions (Murlidhar 2020).

In a different approach, areca fibers were modified with amine-functionalized nanoalumina (F–Al) and later combined with diglycidyl ethers of Bisphenol-A (DGEBA) epoxy resin and made into composites (Fig. 4.2). In this study, diglycidyl ethers of Bisphenol-A (DGEBA) epoxy resin and areca fiber (AF) [80:20] composite, having various weight percentages (1, 2, and 3 wt%) of amine-functionalized nanoalumina (F–Al) were cured at room temperature using polyamidoamine curing agent (XY54). The F–Al acted as an adhesion promoter through its amine groups on the surface, which was confirmed using SEM and TEM. It was observed that addition of 2 wt% F–Al nano reinforcement enhanced the thermal, mechanical, and thermo-mechanical properties of DGEBA/AF nanocomposites (Dhanapal 2022).

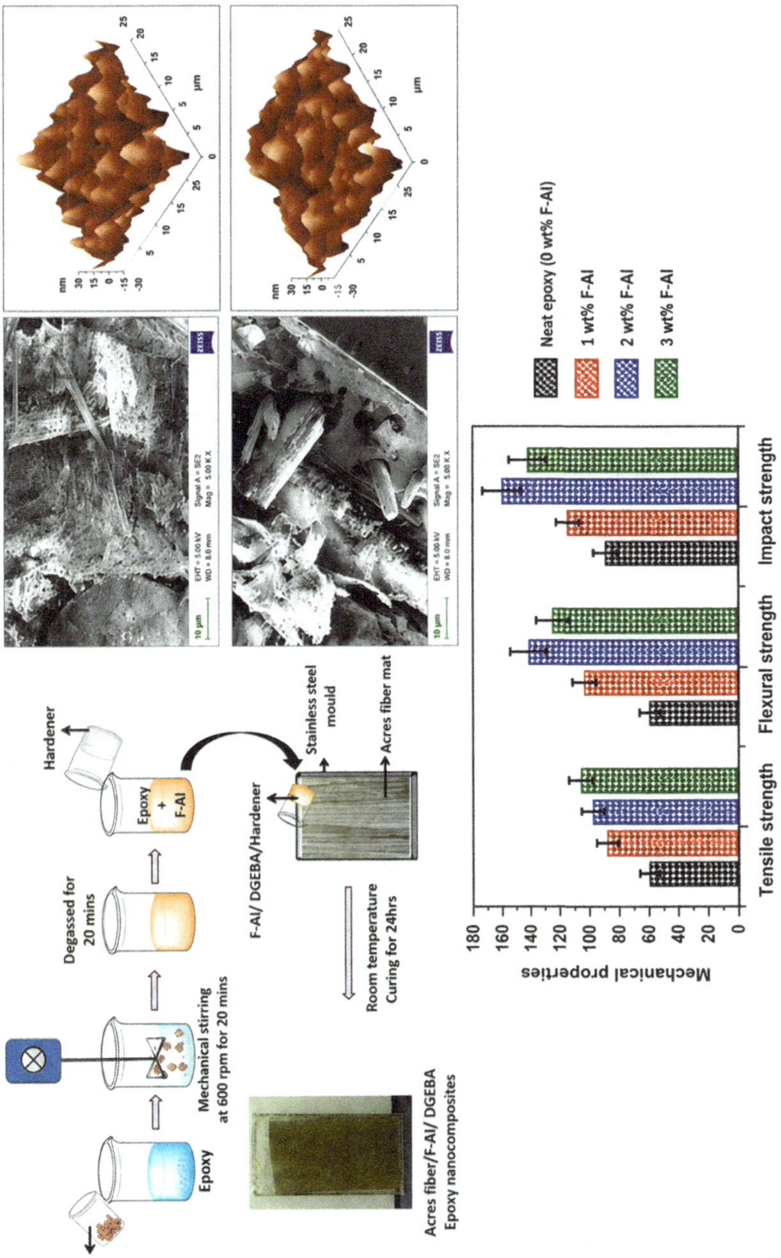

Fig. 4.2 Process of preparing modified areca husk fibers and epoxy reinforced composites (Dhanapal 2022). Reproduced with permission through Elsevier

4.3 Areca-Based Epoxy Composites with Additives

A few studies have been conducted on improving the properties of epoxy-based areca composites with improved properties by using additives and compatibilizers. For instance, a comparative study was done to understand the ability of treated and untreated areca husks and areca nut powder towards improving the properties of epoxy-based composites (Somashekhara 2018) with 5 to 20% areca powder as additive. Composites made using treated areca fibers and 15% areca powder had highest tensile load of 15.6 kN and tensile strength of 23.8 kN (Table 4.3). Similarly, compressive load (24.5 kN) and compressive strength (0.145 kN/mm^2) were also higher compared to neat epoxy. However, the composites containing 5% fibers and 20% powder had lower hardness values compared to 15% fiber and 10% powder composites. Overall, the properties of the composites were exceeding that of commercial Nuwood except in impact strength.

In another study, biosilica prepared from barley husk was treated with 3-aminopropyletriethoxylane to improve the compatibility with the resin (Bisphenol-A type epoxy). Short areca fibers with length between 200 and 300 μm were used as reinforcement (40% w/w) with 1–5% modified barley husk silica as the additive. As seen from Table 4.4, addition of the fibers into the resin increased the mechanical properties, particularly the impact resistance. Similarly, the silica present enhanced tensile and flexural strength up to 3% but decreased at 4%. More than 10 times

Table 4.3 Comparison of the properties of epoxy composites reinforced with untreated and treated areca fibers and with areca nut powder with commercially available Nuwood composites (Somashekhara 2018). Reproduced with permission through open access publication

Composite	Tensile strength, kN	Compressive strength, N	Flexural strength, N	Impact strength, J/mm^2	Hardness in (RHNB)
Untreated areca fiber and areca palm powder					
Untreated F5% P 20%	15.2	24.5	15.7	0.0225	49
Untreated F10% P 15%	15.4	22.0	16.0	0.0225	42.5
Untreated F15% P 10%	15.4	21.8	15.2	0.0243	45.5
Untreated areca fiber and untreated areca palm powder					
Treated F5% P 20%	15.2	21.9	13.8	0.03625	38.5
Treated F10% P 15%	15.6	23.8	16.3	0.05122	37
Treated F15% P 10%	15.5	21.5	15.4	0.03622	39
Commercial Nuwood	2.2	2.0	0.22	0.18\9	2.0

Table 4.4 Properties of epoxy composites reinforced with areca fibers and containing 1 to 4% barley husk ash as additive (Krishnamoorthy 2023). Reproduced with permission from Springer Nature

Composite designations	Tensile strength (MPa)	Flexural strength (MPa)	V-notch rail shear strength (MPa)	Izod impact toughness (Kj/ m^2)	Barcol hardness
RC0	52	86	5.6	0.38	54
RC1	96	106	14.4	3.35	56
RC2	112	142	15.6	4.22	57
RC3	142	174	17.8	4.68	58
RC4	133	148	15.1	4.25	60

increase in toughness and marginal improvement in hardness can be observed (Krishnamoorthy 2023). In a study by Samuel et al., epoxy resin was combined with neem oil (10 or 20%) and used as matrix with silane-treated areca fibers (30%) as reinforcement and 1 or 2% nanosilica as additive, and made into composites at room temperature, which were later cured at 120 °C for 4 h. Addition of neem oil to the epoxy resin marginally decreased the tensile and flexural strength but improved the impact resistance substantially. A nearly 220% increase in tensile strength and flexural strength was observed when areca fibers were added to the resin. Similar improvement was also seen in the Izod impact resistance (Table 4.5). Wear and fatigue resistance also improved due to the addition of the fibers and silica (Samuel 2023). In another study, it was found that reinforcing epoxy with areca fibers with silicon carbide as filler resulted in enhanced tensile and flexural strength up to a filler content of 50%. However, bending and flexural properties were considerably low when less than 50% filler was used, and it was suggested that using 10% filler was optimum to prevent agglomeration. Composites obtained were considered suitable for outdoor applications such as decking, railing, fencing, and siding, etc. (Ramasubbu 2024).

In another approach to improve properties of composites, orange peel was made into biochar and used as an additive for areca fiber reinforced epoxy composites. Fibers having length of 50 mm and diameter of 221 μm, and Bisphenol-A type derived epoxy resin and an aliphatic curing hardener Triethylenetetramine were used to fabricate the composites. Biochar content was varied between 1, 2 or 4% and properties of the composites were compared to composites without biochar and without any fiber reinforcement. As seen from Fig. 4.3, addition of fibers into epoxy increased tensile strength and modulus and inclusion of biochar as additive further increased the tensile strength whereas only marginal improvement was observed in tensile modulus. Similar results were also observed for the flexural properties. However, thermal conductivity increased from 0.224 w/mK for pure epoxy to as high as 0.4 w/mK since the biochar was made of carbon, which supports heat transfer. Dielectric properties also showed an increase with addition of biochar, which makes the composites useful for developing electronic appliances and industrial products (Alshahrani 2024).

Table 4.5 Effect of addition of neem oil, areca fibers, and nanosilica on the mechanical properties of epoxy composites (Samuel 2023). Reproduced with permission from Springer Nature

Composite designations	Tensile strength (MPa)	Flexural strength (MPa)	Izod strength, J	I.L. shear strength, MPa	Shore hardness
Pure epoxy	65 ± 3.25	102 ± 4.5	0.43 ± 0.02	0	86 ± 4.3
Epoxy + 10% oil	65 ± 3.0	92 ± 4.1	0.76 ± 0.03	0	83 ± 3.8
Epoxy + 20% oil	56 ± 2.1	84 ± 3.9	0.92 ± 0.04	0	81 ± 3.2
Epoxy + 20% oil + 30% fibers	124 ± 6.1	188 ± 9.4	5.63 ± 0.28	30 ± 1.5	81 ± 3.1
Epoxy + 20% oil + 30% fibers + 1% silica	134 ± 6.7	205 ± 9.9	6.44 ± 0.35	28 ± 1.3	99 ± 4.9
Epoxy + 20% oil + 30% fibers + 2% silica	148 ± 7.4	233 ± 10.3	6.90 ± 0.46	26 ± 1.2	94 ± 4.4

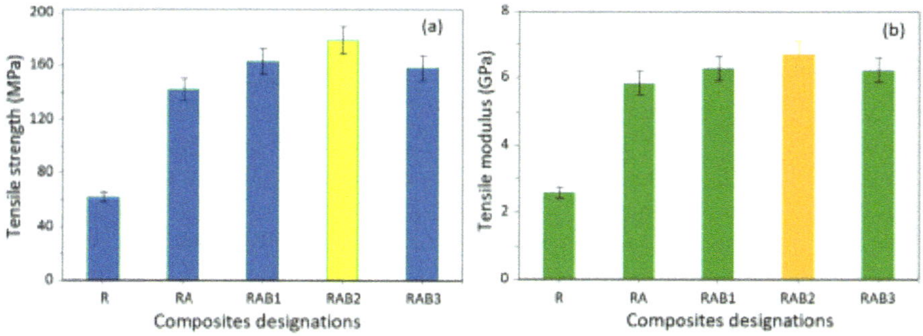

Fig. 4.3 Effect of adding biochar on the tensile strength and modulus of epoxy composites reinforced with areca fibers (Alshahrani 2024). Reproduced with permission from Springer Nature

In a related study, biochar was prepared from papaya fruit (edible part) (Fig. 4.4) and combined with short areca fibers (50 mm length), which were both subject to silane treatment. Composites were manufactured using epoxy resin and 0 to 4% biochar as the additive (Kishore 2022). Neat epoxy had tensile strength and modulus of 64 and 2588 MPa, respectively, which increased to as high as 178 and 6722 MPa with the addition of 2% biochar. Although only a modest increase was seen in flexural strength, the flexural modulus increased to 6835 MPa. Impact resistance had more than 400% increase from 0.42 J up to as high as 5.42 J. With good thermal conductivity of 0.426 W/mK and dielectric constant of 3.8, the composites were considered to be suitable for various electronic applications (Kishore 2022).

Fig. 4.4 Process of preparing the papaya fruit powder for carbonization and use as additive in composites (Kishore 2022). Reproduced with permission through open access publishing

Instead of using non-biodegradable additives such as silica, epoxy resin was reinforced with areca fibers (10–40%), and 1 to 3% coconut shell powder was used as additive. Composites containing 30% fibers and 1% coconut shell powder had maximum tensile strength and modulus, flexural strength and modulus, and also interlaminar shear strength (Chowdari 2021). Unlike most researchers who use one additive in composites, Shunmugasundaram et al. adopted vacuum infusion technique to add two nanofillers, graphene and multiwalled carbon nanotubes, into areca fiber reinforced epoxy composites. About 7% improvement in tensile strength and 22% increase in flexural strength were observed (Shunmugasundaram 2020). Al_2O_3 was used as a filler for areca husk reinforced epoxy composites with considerable improvement in properties, including tensile, impact, and hardness resistance. Shore hardness increased from 73.6 to 81 with the addition of 8% Al_2O_3. Similarly, there was about 37% increase in tensile strength from 16.8 to 23 MPa and nearly 3 times increase in impact resistance from 59 to 180 J/m^2. Fourier transform infrared (FTIR) measurements confirmed physical and chemical interactions between the matrix and fibers through hydrogen bonding, van der Waals forces, etc., leading to better stress transfer and improved composite properties (Tavadi 2025).

4.4 Hybrid Composites with Epoxy as Matrix

In addition to using areca husks or fibers as reinforcement for epoxy with and without additives, several studies have been conducted to develop hybrid composites using areca fibers with other natural fibers and biomass, and also with synthetic polymer-based materials. In one such attempt, hybrid epoxy composites were developed by combining areca fibers with fibers obtained from desert cotton (0.25–0.5 cm in length and 2–3 μm in diameter). Images of the various composites formed are shown in Fig. 4.5, and some of the mechanical and thermal properties of the composites are listed in Table 4.6. The composites obtained had excellent sound absorption potential with sound absorption coefficient of up to 0.7 at 6400 Hz. High electrical surface resistivity and electrical volume resistivity were also observed for the hybrid composites, making them suitable for various construction applications (Srinivasan 2023). In another study on developing epoxy composites reinforced with areca fibers, the husks were initially subject to chemical extraction (6% by volume NaOH alkali solution for 24 h at room temperature of 26 ± 2 °C) when fine and coarse fibers were obtained. The fine and coarse fibers were combined with the epoxy resin HSC 7600 (1.16 g/cc) and epoxy hardener HSC 8210 (0.93 g/cc) along with glass fibers, and made into hybrid composites. Tensile properties of the composites varied considerably depending on the fiber type used. Highest tensile strength (24.8 MPa) was obtained for the hybrid composites containing coarser fibers, whereas finer fibers provided strength and modulus of 15.1 and 3.2 GPa and 10.8 and 3.5 GPa, respectively (Murlidhar 2019). Similarly, flexural strength varied between 73, 66.7, and 284 MPa.

In another study, reinforcing epoxy with areca fibers in various orientations with glass fibers was shown to improve the tensile behavior and other performance properties of composites (Babu 2019). Alkali-treated areca and sunn hemp fibers were combined with uniaxial or biaxial glass fiber fabrics (360 g m^{-2}, diameter of 5–10 μm) and used as reinforcement for epoxy resin (LY556), and having room temperature curing agent, polyamine hardener (HY951). Fiber content for composites without glass fibers was 60%, and those with glass fibers (10%) were maintained at 40%. Depending on the fiber content and type of glass fibers used, the maximum load varied from 24 to 148 N and maximum flexural stress was between 28 and 271 MPa. The composites developed were considered suitable for false ceilings, partitions, interiors, and sports equipment (Thamba 2019).

Using epoxy as resin, biocomposites were developed with areca and banana fibers as reinforcement. Areca fibers were treated with 5% alkali and later with silane (4%) to improve adhesion and other properties. Composites were developed using hand lay-up techniques with various layers of banana and areca fibers in different sequences. Tensile properties, particularly tensile modulus, showed considerable increase when only areca fibers were used, but impact strength was highest (4.1 J/m) for banana fiber reinforced composites. The composites were found to be useful for lightweight applications (Dhilip 2024). In another study, areca fibers were blended with ramie fibers, and graphene nanoparticles (0.5–2%) were included as additive

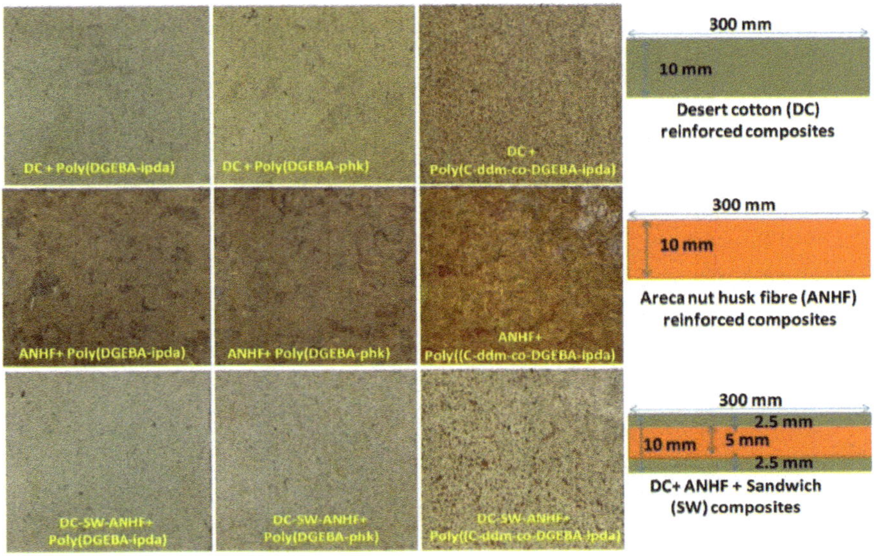

Fig. 4.5 Desert cotton (DC) and areca nut husk fibre (ANHF) reinforced epoxy and benzoxazine/epoxy composites panels (Srinivasan 2023). Reproduced with permission from Elsevier

for epoxy. About 188% higher tensile strength, 143% higher flexural strength, and 160% higher impact energy were observed due to the addition of 1.5% graphene. Composites developed were considered to be suitable for weight-sensitive engineering applications (Dhilipkumar 2024). Similarly, various ratios of areca and sisal fibers were blended together and made into composites with epoxy as the resin. Areca fibers were found to provide better strength than sisal fibers and the composites had good mechanical properties and stability to water (Prabakaran 2017).

Using a similar approach, hybrid composites were developed using areca fiber fabrics, sisal fiber fabrics, and carbon fabrics in various configurations with epoxy as the matrix. Carbon fiber reinforced epoxy had substantially higher tensile strength compared to those reinforced with only sisal or areca fibers. However, hybrid composites containing carbon as the outer layer provided highest flexural strength of 692 MPa and interlaminar shear strength (ILSS). Hence, the properties of the composites could be controlled by varying the amount and architecture of the various layers. Composites obtained were considered suitable for use in railway interiors and other applications (Jagadeesh 2022). To develop epoxy-based hybrid composites, epoxy resin having density of 1.18 g/cm^3 was combined with silane-modified chitosan and silane-treated areca fiber mats with a density of 1.05–1.25 g/cm^3. Further, these materials were sandwiched and layered between goat leather having thickness of 0.3 to 0.6 mm. Process of developing the hybrid composites is shown in Fig. 4.6. Mechanical properties of the composites showed substantial differences depending on the materials used and arrangement of the reinforcement in the different layers. Maximum tensile strength of 172 MPa, modulus of 4.98 GPa, and flexural strength

Table 4.6 Thermal and mechanical behavior of areca husk/fiber reinforced epoxy-based composites (Srinivasan 2023). Reproduced with permission from Elsevier

Sample name	Mechanical behavior			
	Tensile strength (±0.5 MPa)	Tensile modulus (±10 MPa)	Flexural stress (±0.5 MPa)	Flexural modulus (±10 MPa)
Neat polymer matrix				
Poly(DGEBA-ipda)	45	2565	74	1716
Poly(DGEBA-phk)	46	1832	85	1861
Poly(C-ddm-*co*-DGEBA-ipda)	48	1865	97	2312
Desert cotton (DC) reinforced composites				
DC/poly(DGEBA-ipda)	2	327	3	165
DC/poly(DGEBA-phk)	2	258	3	226
DC/poly(C-ddm-*co*-DGEBA-ipda)	3	293	4	187
Areca nut husk fibre (ANHF) reinforced composites				
ANHF/poly(DGEBA-ipda)	7	300	7	622
ANHF/poly(DGEBA-phk)	7	425	9	96
ANHF/(C-ddm-*co*-DGEBA-ipda)	8	413	10	520
DC-sw-ANHF reinforced composites sandwich				
DC-sw-ANHF/poly(DGEBA-ipda)	6	492	6	262
DC-sw-ANHF/poly(DGEBA-phk)	6	495	7	374
DC-sw-ANHF/poly(C-ddm-*co*-DGEBA-ipda)	6	602	8	793

and modulus of 208 MPa and 5.3 GPa were obtained when 1% chitosan was added into the composites. Silane treatment improved the fatigue life counts to 32,847 and also enhanced the load-bearing capacity of the composites (Surianarayanan 2024).

In a related study, epoxy resin having density of 1.1–1.4 g/cm^3, elastic modulus between 3 and 6 GPa and tensile strength between 30 and 100 MPa was reinforced with a blend of areca, snake grass, and kenaf fibers which were used in the raw form and also after treating with water and alkali. The overall fiber content in the composites was maintained at 40% and resin content was 60%. Alkali-treated fibers provided better strength, and a combination of 10% areca, 10% kenaf, and 20% snake grass fibers provided the highest tensile and flexural strength, impact resistance, and hardness values. Alkali treatment enhanced fiber–matrix adhesion, and no major fiber pull out was observed in the scanning electron microscopy (SEM) images (Rangaraj 2022). To develop hybrid composites, fibers extracted from areca husks were combined with bagasse in several layers, and an epoxy resin and hardener was made to flow between the layers. The materials were compressed at 50 kN at

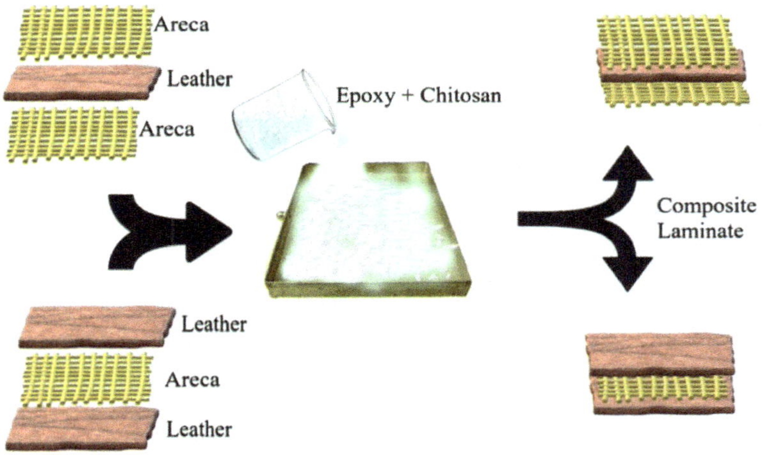

Fig. 4.6 Graphical representation of the materials and processes used to develop the hybrid composites (Surianarayanan 2024). Reproduced with permission from Springer Nature

room temperature to form the composite panel intended for false roofing applications. Composites were made using three different areca fiber contents (13.6, 16.9, and 19.7%). No major difference was observed in the tensile strength of the different samples but flexural strength was highest for composites containing 19% bagasse. However, thermal conductivity was higher for the composites with higher fiber content (Mahmud 2025).

4.5 Polypropylene (PP) as Matrix

Polypropylene is a common thermoplastic polymer extensively used as matrix for biocomposites because of its relatively low (165–190 °C) melting temperature. PP has been used in pellet, fiber, liquid, non-woven, and various other forms as matrix for composites. To develop PP based composites, areca fibers were subject to treatment with permanganate ions, acrylic acid or benzoylation and the treated fibers were used as reinforcement for PP composites. Chemical treatments reduced the hydrophilicity, improved adhesion between matrix and reinforcement leading to substantially enhanced tensile properties compared to composites that were developed only using the untreated fibers (Dhanalakshmi 2015). For instance, flexural strength of the composites had increased to about 50 N/mm^2 for the acrylic acid-treated fibers compared to 45 N/mm^2 for the composites made using the untreated fibers (Fig. 4.7).

Since PP is hydrophobic and areca fibers are hydrophilic, it is common to use a compatibilizer to improve compatibility and performance properties. In one such study, PP was reinforced with 20–50% areca fibers, and MAPP (maleic anhydride

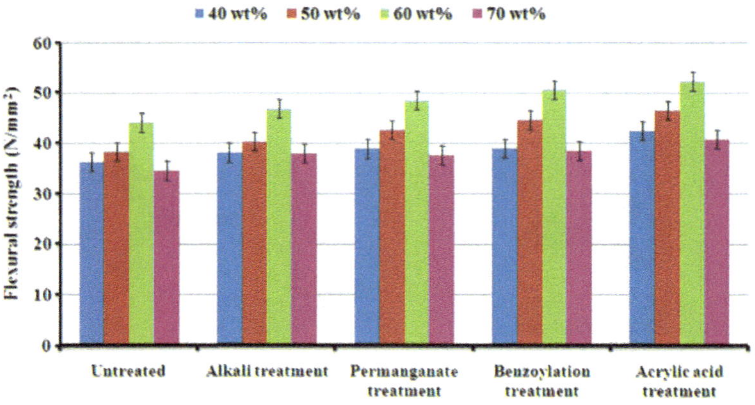

Fig. 4.7 Effect of various chemical treatments of areca fibers on the flexural strength of the PP composites (Dhanalakshmi 2017). Reproduced with permission through open access publishing

grafted polypropylene) was added as the compatibilizer. Addition of the compatibilizer improved the tensile properties due to enhanced interfacial adhesion, which enabled efficient transfer of load from the matrix to the fibers. Considerable reduction in water sorption of the composites was also noticed making them suitable as replacement for natural wood (Venkatachalaiah 2022). Further, the MAPP compatibilized areca husk fiber reinforced PP composites were studied for their electrical and thermal behavior at various proportions of the areca fibers. Dielectric constant of the composites varied considerably with change in fiber content and was found to be 550 at a frequency of 10 Hz. However, impedance values decreased with increasing fiber content whereas considerable increase was noticed in the electrical conductivity. Overall, these composites, having high dielectric constants, were considered suitable for potential applications in gate dielectrics, actuators, sensors, and others (Venkatachalaiah 2025). In another study, areca fibers obtained through natural water retting were used as reinforcement for polypropylene, and it was observed that addition of 10% fibers improved the tensile, impact, and bending strength to the highest extent. Further improvement in the properties of the composites and enhanced biodegradability were obtained when 10% sea weed powder was included as an additive (Hassan 2010).

In an attempt to improve recycling and reduce environmental pollution, hybrid composites were developed by combing discarded nylon, polypropylene, and areca husk fibers. The fibers were extracted from areca husks using natural retting and later subject to alkali treatment (5% NaOH, 70 °C for 2.5 h) and combined with the matrix polymers and compression molded into composites. Fiber content was varied between 10 and 25%, and compression was done at temperature between 160 and 185 °C for 12–15 min. There was a gradual increase in properties of the fibers with increasing fiber content from 5 to 20% but no major difference was observed in tensile properties when treated or untreated fibers were used. Both areca and nylon fibers were considered to be suitable to improve the properties of PP composites and help in reducing waste and developing biobased materials (Sidikka 2015, 2022). In a similar approach, hybrid composites were developed by combining areca and banana fibers in a 1:1 ratio with PP as the matrix. The fiber content in the composites was varied between 5 and 15%, and composites were compression molded at 160 °C, 30 kN pressure for 15 to 20 min. Addition of the fibers increased tensile, flexural properties, and impact resistance, but the extent varied depending on the amount of fibers. Composites containing 10% fibers were found to provide the highest thermal resistance (Akter 2018).

In yet another attempt at developing areca fiber-based hybrid composites, areca husks were soaked in water for 15 days and later treated with 10% NaOH for 30 min at 95 °C (Fig. 4.8). Similar treatment was also done to obtain banana fibers. The areca and banana fibers were combined in various proportions and used as reinforcement for polypropylene with fiber content varying from 40 to 60%. Both tensile strength and bending strength, and tensile elongation were higher for the 60% fiber containing composites. Water absorption of the composites also increased from 2.25 to 3.75% when the fiber content was increased from 40 to 60% (Mollah 2024).

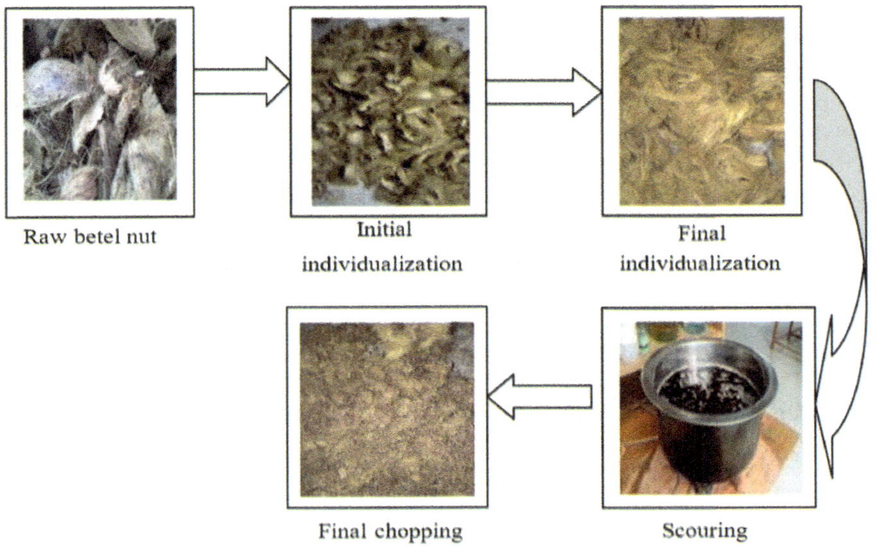

Fig. 4.8 Process of extraction of fibers from raw areca (betel) nut husks (Mollah 2024). Reproduced with permission through open access publication

4.6 Formaldehyde as Resin

Although considered to be toxic and carcinogenic, formaldehyde has been used as matrix for biobased composites, including areca husks. In a simple approach, areca husks were soaked in 5, 10, 15, 20, and 25% NaOH solution for about 12, 24, 36, 48, and 60 h, and the fibers obtained were used as reinforcement with phenol formaldehyde as the binder/matrix. Fiber content in the composites was varied from 10 to 50%, and composites were formed by compression molding at 140 °C and 2 MPa pressure for 30 min. Composites developed were able to resist tensile load of up to 622 N and bending load of 223 N with good resistance to moisture, making them suitable for packaging and structural applications (Swamy 2004). In another study, areca husks were soaked with water, dried, and later treated with 10% potassium hydroxide solution for 36 h at 30 °C. Both the chemically treated and untreated fibers were used as reinforcement for urea formaldehyde and melamine formaldehyde resins in either 50 or 60 w/w% (Srinivasa 2011a, b). Properties of the composites were significantly affected based on the use of treated or untreated fibers, with the chemically treated fibers providing better adhesion between fibers and matrix due to enhanced polar attractions, leading to better properties. In addition to the proportion of the fibers, the curing conditions were found to affect the impact strength and hardness of the composites (Srinivasa 2011b).

A similar approach to treat the husks and obtain fibers was used to develop composites using different resins such as urea formaldehyde, melamine urea formaldehyde and epoxy with 50 to 60% fibers as reinforcement was reported by Srinivasa et al. (2011) Epoxy provided the highest flexural, impact and bending strength compared to the other two resins and composites were considered suitable for furniture, packaging and as wood replacements. Areca fine fibers having diameter between 0.285 and 0.89 mm and length between 18 and 40 mm were used as reinforcement for phenol formaldehyde. In addition, various concentrations of silicon carbide and alumina oxide were added to improve performance properties. As seen from Table 4.7, flexural strength and modulus showed improvement with the addition of fibers, Al_2O_3, and SiC. Similar improvement was also observed for tensile strength, which was dependent on the size of particles used as well (Raja 2020).

4.7 Unsaturated Polyester Resin

In an effort to find the most optimum properties of areca husk fiber reinforced composites, short areca fibers were combined with unsaturated polyester resin, methyl ethyl ketone peroxide, and cobalt napthanate (98:1:1 ratio) and made into composites measuring 300 × 150 × 3 mm. Fiber content in the composites was varied from 10 to 50% using random orientation. Fiber content of up to 40% improved tensile strength, but debonding and fiber pull-out were noticed above a fiber content of 40%. Highest tensile strength obtained was 68 MPa, which was similar to that of glass fiber

Table 4.7 Variations of flexural properties of areca fiber reinforced phenol formaldehyde composites with various levels of alumina oxide or silicon carbide as additives (Raja 2020)

Particle content, %	Alumina oxide		SiC	
	Flexural strength, MPa	Flexural modulus, MPa	Flexural strength, MPa	Flexural modulus, MPa
No reinforcement	36.9	1214	36.8	1214
40AFF/PF	64.6	1357	64.6	1357
2.5	69.5	1382	73.5	1397
5.0	74.1	1428	79.7	1442
7.5	81.7	1486	85.9	1498
10.0	76.7	1412	80.1	1468

reinforced composites (72 MPa) and better than composites reinforced with other natural fibers (Binoj 2016; Binoj 2022 a, b) (Table 4.8). Tribological tests revealed that the composites containing the fibers had lower specific wear rate and coefficient of friction (Binoj 2023). Areca husk fibers were specifically considered to be suitable for wear applications because of the various advantages. Several reports are available on the development and performance of areca husk reinforced composites for tribological applications, which have been summarized in Table 4.9 (Raj 2022).

Alkali (potassium hydroxide, 10%) treated areca husk fibers were combined with polyester resin in 40 or 50 wt% ratio and made into composites. Impact resistance and hardness properties were tested and found to improve due to the incorporation of the fibers. Rockwell hardness numbers ranged from 20 to 35, depending on the proportion of the fibers and conditions during manufacture (Madu 2019). In fact, adding alkali-treated and untreated areca husk fibers into polyester resin provided composites with tensile, flexural, and impact resistance, considerably higher than plywood of the same thickness, with composites having tensile strength of 18 MPa, flexural strength of 34 MPa, and impact strength of 1.62 kJ/m^2. Interestingly, the composites were able to maintain these properties even under wet conditions (Islam 2018). Fibers were extracted from areca husks by soaking in water for 5 days, and the obtained fibers were further treated with 10% potassium hydroxide for 36 h at 30 °C. Composites were obtained by combining polyester resin with 50 or 60% fibers and casting in a mold to the desired shape and size. Final curing of the composites was done for 360 to 1080 h. Properties of the composites were found to be highly dependent on the ratio of the fibers and curing temperature (Madu 2019). Fibers were extracted from areca husks by soaking in water for 5 days, and the obtained fibers were further treated with 10% potassium hydroxide for 36 h at 30 °C. Composites were obtained by combining polyester resin with 50 or 60% of fibers and casting in a mold to the desired shape and size. Final curing of the composites was done for 360 to 1080 h. The properties of the composites were found to be highly dependent on the ratio of the fibers and the curing temperature (Madu 2019).

Table 4.8 Comparison of the properties of areca husk fiber reinforced composites with glass fiber reinforced polyester resin composites (Binoj 2022)

Reinforcement	Fiber, wt %	Tensile strength, MPa	Tensile modulus, GPa	Tensile strain, %	Flexural strength, MPa	Flexural modulus, GPa	Flexural strain,%	Impact strength, J/ cm^3	Hardness (HRRW)
0	0	37.0	1.5	0.02	41.00	1.80	0.02	0.63	64
Areca fiber	10	38.07	0.96	3.93	41.46	0.98	4.23	3.57	63
	20	48.52	1.11	4.37	52.36	1.21	4.31	4.34	67
	30	57.61	1.23	4.67	66.59	1.52	4.36	5.27	70
	40	68.20	1.32	5.16	73.91	1.62	4.56	6.82	77
	50	52.18	1.15	4.53	63.84	1.47	4.34	5.15	67
Glass fiber	10	46.8	23.4	2.00	32.07	11.15	2.88	7.33	72
	20	51.1	24.3	2.10	33.83	8.72	3.88	7.47	76
	30	52.1	23.7	2.20	40.10	9.69	4.14	7.6	79
	40	72.1	24.9	2.90	53.17	12.11	4.39	7.68	88
	50	63.1	24.6	2.57	47.13	11.25	4.19	7.52	74

Table 4.9 Comparison of the wear performance of areca husk fiber reinforced composites developed using various resins and treatment conditions (Raj 2022)

Materials	Method of fabrication of composite	Wear testing machine	Conditions for wear testing	Predominated wear mechanism observed in SEM	Results
Unsaturated polyester + Betel fiber	Hand lay up	Recently developed block-on-disc machine	Load applied (5–30 N), sliding distances (0–6.72 km), dry contact conditions sliding velocity-2.8 m/s	Plastic deformation, debonding of fibers, micro- and macro-cracks	Higher wear performance during parallel orientation of fiber mat
Unsaturated Polyester(Butanox M 60) + Betel fiber	Hand lay up	Pin on disc	Load applied(5–30 N), and sliding distances (0–6.72 km), dry/wet contact conditions, sliding velocity-2.8 m/s	Wet contact conditions debonding, delamination and detachment of fibers	6% NaOH fiber treatment enhances wear resistance under dry/wet contact conditions
Butanox M-60 methyl ethyl ketone peroxide + Betel fiber	Hand lay up	Pin on disc	Load applied (5–30 N), sliding distances (0–6.72 km) at sliding speed of 2.8 m/s	Plastic deformation, debonding of fibres, Micro-cracks	Wear performance of the composite demonstrated rising tendency of friction coefficients at various applied loads (5–30 N)
Betel fiber and groundnut shell powder + PLA resin	Hand lay up	Pin on disc	Sliding velocity 2 m/s, Applied load 5, 10, 15, 20 and 25 N	No SEM observations were done	The presence of betel fiber and ground nut shell powder decreases the wear rate of composite

(continued)

Table 4.9 (continued)

Materials	Method of fabrication of composite	Wear testing machine	Conditions for wear testing	Predominated wear mechanism observed in SEM	Results
Betel nut fiber mat + Polyester resin	Hand lay up	Block-on-disc	Sliding distances (0 – 6.72 km) and load applied (20 – 200 N) at 2.8 m/s sliding velocity, wet conditions and different fiber mat orientations	Debonding of fiber	Fiber mat in normal orientation shows superior wear performance
Epoxy + Treated betel fiber	Hand lay up	Developed linear tribo machine (LTM)	Abrasive particle dimensions (500, 714 and 1430 lm) and sliding velocities (0.026–0.115 m s1) at constant applied load (5 N) three-body abrasion	Pitting and pullout of betel nut fibres plastic deformation	The abrasive wear depends on the abrasive particles dimensions and sliding velocity and Coarse sand shows better frictional coefficient

In addition to the reinforcement and matrix, additives are included to improve properties. In such an attempt, composites were developed using alkali-treated areca husk fibers and unsaturated polyester resin as matrix with 1 to 4% (w/w%) of silicon carbide as filler. It was found that addition of 3% filler provided considerably high improvement with composites having properties desirable for marine and lightweight applications (Brailson 2023). Using unsaturated polyester (UPE) as resin and methyl ethyl ketone peroxide (MEKP) as hardener and cobalt napthanate as the catalyst, composite were prepared with alkali or potassium permanganate-treated areca husk fibers as reinforcement (Binoj 2023). Chemically treated fibers improved thermal stability and viscoelastic behavior, and also enhanced bonding strength between fibers and matrix. SEM images (Fig. 4.9) showed that composites had a rough surface, and no fiber pull outs were observed.

Since coir and areca belong to the same family and both are obtained as byproducts and have similar composition and structure, several studies have developed hybrid composites by combining areca and coir fibers. Areca and coir fibers were separately treated with 10% NaOH for 30 h at room temperature and later used to develop composites containing 20% coir, areca, and their blends with unsaturated polyester (UPE) resin as the matrix. Interestingly, areca fiber reinforced composites provided highest flexural modulus, and coir fiber composites the highest tensile

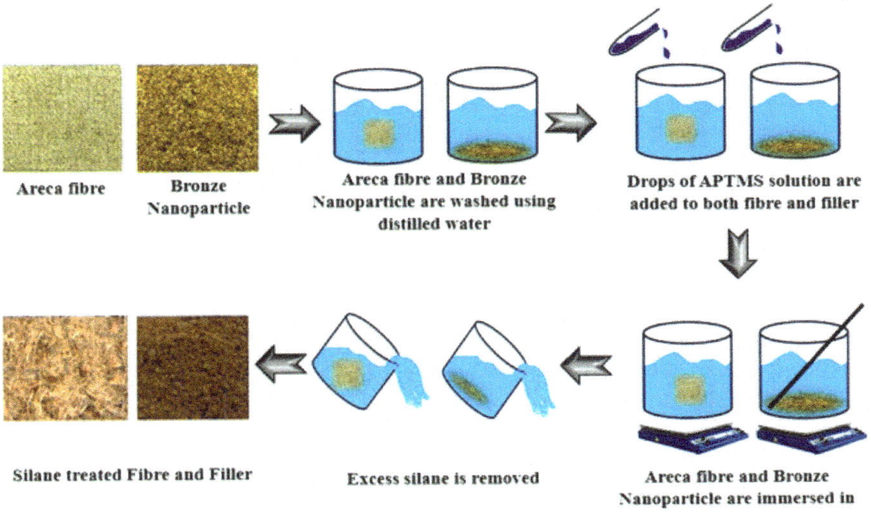

Fig. 4.9 Process used to prepare the areca fibers and bronze nanoparticles used as reinforcement and filler, respectively (Balguri 2024). Reproduced with permission from Springer Nature

strength. However, blending of the two fibers did not provide higher properties compared to the composites containing the individual fibers (Padmaraj 2018). In another study, areca fibers and coir were treated with 10% NaOH for 30 min and used as reinforcement for unsaturated polyester resin. After treatment, areca husk fibers had density of 0.9 g/cm^3, coir fibers were of 1.12 g/cm^3, and polyester resin was 1.13 g/cm^3. Properties of the composites were dependent on the type of fiber used. Flexural strength and impact resistance were higher for areca and hybrid fiber reinforced composites, whereas tensile strength was marginally higher for the coir reinforced composites, which should be due to the inherent properties of the fibers (Padmaraj 2018). In another study, cenospheres were used as fillers for composites containing untreated and alkali-treated areca husk fibers as reinforcement and UPE as resin. Good bonding was noticed between the matrix and reinforcement, and the composites obtained had good thermal insulation and improved hardness.

Biochar made from vegetable waste was added to composites made from areca fibers and unsaturated polyester resin. The resin used had density of 1.13 g/cc, viscosity of 6000 cps, and molecular weight of 2400 g/mol. Fibers used had diameters of 150, 200, and 250 μm and length between 20 and 40 mm. Composites were made using methyl-ethylketone peroxide as hardener and cobalt naphthenate as accelerator. Rebars made from the composites had tensile strength of 50 MPa and low water absorption of 1.27% considered suitable for various construction applications including manhole covers (Kumar, 2024). Several researchers have attempted to use various fillers to improve the properties of areca fiber reinforced composites. For example, biosilica was extracted from wheat husks and combined with areca husk fibers and polyester resin in various proportions (0.5 to 4%) (Thilagam 2024). Silane

Table 4.10 Properties of the composites made from areca husk fibers and polyester resin with and without biosilica as the additive (Thilagam 2024). Reproduced with permission from Springer Nature

Composite designations	Tensile strength (MPa)	Tensile modulus (GPa)	Flexural strength (MPa)	Flexural modulus (GPa)	Izod impact (J)	Hardness (Shore-D)
UP	47 ± 2.4	1.48 ± 0.2	66 ± 2.2	2.36 ± 1.2	0.38 ± 0.2	77 ± 1.2
UPA	72 ± 2.2	2.74 ± 0.6	102 ± 2.5	3.82 ± 0.4	3.84 ± 0.5	78 ± 1.5
UPA1	94 ± 2.7	3.28 ± 0.4	128 ± 2.7	3.27 ± 0.5	4.38 ± 0.8	80 ± 1.7
UPA2	116 ± 2.1	3.81 ± 0.5	141 ± 2.5	4.93 ± 0.4	4.66 ± 0.7	82 ± 1.4
UPA3	137 ± 2.3	4.69 ± 0.7	177 ± 2.8	5.37 ± 0.3	5.14 ± 0.5	85 ± 1.9
UPA4	102 ± 2.1	3.53 ± 0.3	152 ± 2.4	5.11 ± 0.2	5.84 ± 0.3	86 ± 1.6

treatment was also done on the fibers to improve compatibility between the fibers and the matrix. Some of the mechanical properties of the composites are given in Table 4.10. Addition of 2% silica provided the maximum improvement in mechanical properties, as seen from the Table (Chethan 2023; Thilagam 2024).

Red mud composed of 15.2% SiO_2, 16.8 Al_2O_3, 34% Fe_2O_3, and 11.9 Na_2O_3 was used as filler for areca fiber reinforced polyester resin. Size of the red mud particles ranged between <0.075 mm and >1.2 mm. Fiber content in the composites was varied between 5 and 30%. Compressive strength of the composites varied considerably depending on the fiber and filler content, with highest compressive stress of 49 MPa (Boopathi 2019). The red mud particles improved the binding ability of the matrix and fibers. In addition to using untreated and alkali-treated areca fibers, silane-treated areca fibers were considered as reinforcement for polyester resin with bronze nanoparticles (also silane-treated) as additives. Fiber content in the composites was maintained at 40%, and the amount of nanoparticles was varied from 1, 3 and 5%. Mechanical properties, including tensile, flexural, impact, and hardness improved with the addition of the nanoparticles. A low thermal conductivity of 0.212 W/mK and water sorption of 0.12% provided good hydrophobicity and made the composites suitable for automotive, aviation, shipbuilding, and other applications (Balguri 2024).

4.8 Vinyl Resin as Matrix for Areca Husk Fibers

Areca husk fibers were obtained from unripe, ripe, and matured husks with varying physical properties obtained after treating with 5% NaOH for 30 min. These fibers were combined with a vinyl resin (EPOVIA RF-1001) and 1% methyl ethyl ketone peroxide (MEKP). Composites were obtained by combining various ratios of the fibers with the resin. Both debonding and interfacial shear stress (IFSS) were higher for the fibers obtained from the mature husks. However, after alkali treatment, the fibers obtained from the ripe husks had considerably higher properties (Table 4.11)

Table 4.11 Comparison of the IFSS and debonding force of composites reinforced with areca husk fibers (Lazim 2014). Reproduced with permission through open access publishing

Type of fibre		Debonding force (N)	IFSS (MPa)
Untreated	Unripe	5.95 ± 0.27	1.34 ± 0.03
	Ripe	5.22 ± 0.06	1.23 ± 0.11
	Matured	6.08 ± 0.05	1.57 ± 0.02
Alkali treated	Unripe	8.71 ± 0.47	1.96 ± 0.08
	Ripe	14.16 ± 0.39	3.34 ± 0.14
	Matured	7.14 ± 0.21	1.84 ± 0.06

(Lazim 2014). Addition of fibers from unripe, ripe, or matured fibers provided considerably different properties with lower flexural strength but higher flexural modulus (Fig. 4.10) compared to unreinforced vinyl ester. However, both tensile strength and modulus were higher for the reinforced composites with ripe fibers providing highest strength and matured fibers the highest modulus. In terms of fiber content, 10% fiber reinforcement showed the most improvement in properties, and at higher fiber content, whereas there was considerable decrease in properties mainly due to incompatibility and uneven distribution of fibers, but the moisture sorption of the composites also had considerable increase (Yusriah 2018). In another study, lightweight vinyl ester-based composites intended for prosthetic applications were developed using silane-treated areca fiber and silane-treated PET foam. Mechanical properties, particularly interlaminar shear strength increased substantially without any delamination. Improvements were also observed in flame resistance, thermal conductivity, and resistance to water immersion were observed even after accelerated aging at 40 and 50 °C (Khan 2024).

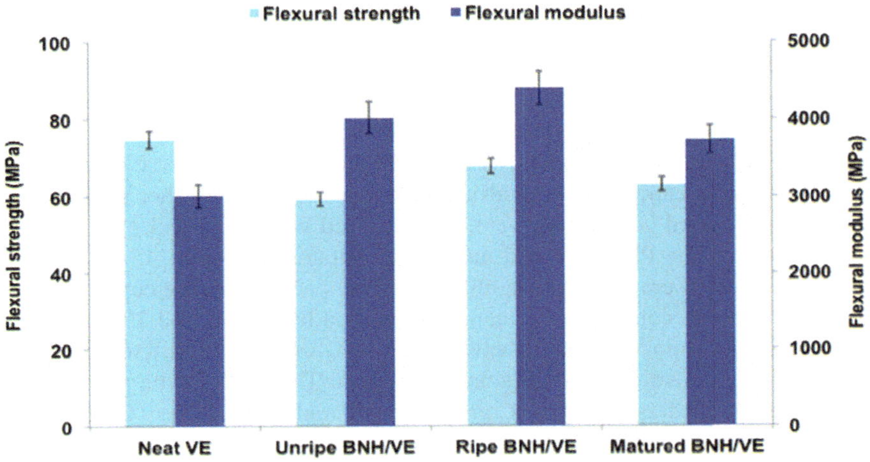

Fig. 4.10 Flexural strength and flexural modulus of neat vinyl ester (VE) and VE composites reinforced with 10 wt.% unripe, ripe, and matured BNH fiber (Yusriah 2018). Reproduced with permission from Elsevier

4.9 PET Matrix

Polyethylene terephthalate (PET) is another common synthetic matrix used for composites. For instance, areca fibers (diameter of 28 to 46 μm and a staple length of 40 mm) were made into non-woven mats and later used as reinforcement for poly(ethylene) terephthalate fibres of 1.4 denier, staple length of 38 mm, and a thermal bonding temperature of 115 °C. Panels were made with up to 30% fiber content, which provided thermal resistance between 0.062 and 0.100 m^2K/W and sound absorption coefficient of 0.35 dB at 6300 Hz making them suitable for building insulation applications (Muthukumar 2023). In another study, areca husks were treated with 5% NaOH at 80 °C for 1 h, and fibers obtained were cut into lengths of 3cm. Later, the fibers (20%) were combined with polyethylene terephthalate (PET) and compounded at 260 °C. Composites were obtained with the desired dimensions suitable for testing. Composites obtained had a microhardness of 27.7 and tensile strength of 57 MPa, flexural strength of 53 MPa, and impact strength of 22.5 kJ/m^2. It was observed that these properties were sufficient for most composite applications and that chemical treatment of the husks was necessary to obtain better properties (Nayak 2022).

A combination of areca fiber mats, PET foam core (recycled), and cellulose modified vinyl ester resin was used to develop composites. The areca fibers and PET foam were silane-treated to improve compatibility with the matrix, where the PET foam was placed between the areca fibers. The composites developed were subject to accelerated water and thermal aging. Properties of the composites varied depending on the composition and conditions during aging. Combing PET core with silane-treated areca fibers provided considerably enhanced mechanical properties, thermal conductivity (Fig. 4.11), and fire resistance, making the composites suitable for housing interior, automotive, aviation, electronic goods, and other applications (Rajeshkannan 2024).

To develop hybrid composites, areca husk fibers and bagasse were blended together and used as reinforcement for polyethylene. However, composites made using the blend (50/50%) fibers had considerably lower strength (12.1 MPa) compared to using only areca fibers (28.1 MPa) or only bagasse fibers (34.8 MPa) as reinforcement. Areca fibers provided better strength than bagasse when treated with higher concentrations of alkali (Herlinawati 2022). In another study, hybrid polyethylene-based composites were manufactured with areca and glass fibers as reinforcement. The PET used had melting temperature between 115 and 140 °C, and areca husks were treated with alkali and fibers obtained were cut into lengths of 3–4 mm. Fiber content in the matrix was varied between 5 and 20% and areca and glass fiber ratio was varied between 1:1, 3:1, or 1:3 with a fixed fiber ratio of 20%. To improve compatibility and properties, TiO_2 and zeolite powder (2%) were also included as additives. Composites containing 5% fibers (1:1 ratio) had highest tensile strength whereas modulus was highest for composites containing 20% fibers. However, flexural strength and modulus, hardness and impact strength improved continually as amount of fibers increased, and incorporating the additives only provided a marginal improvement in properties of composites (Haque 2013).

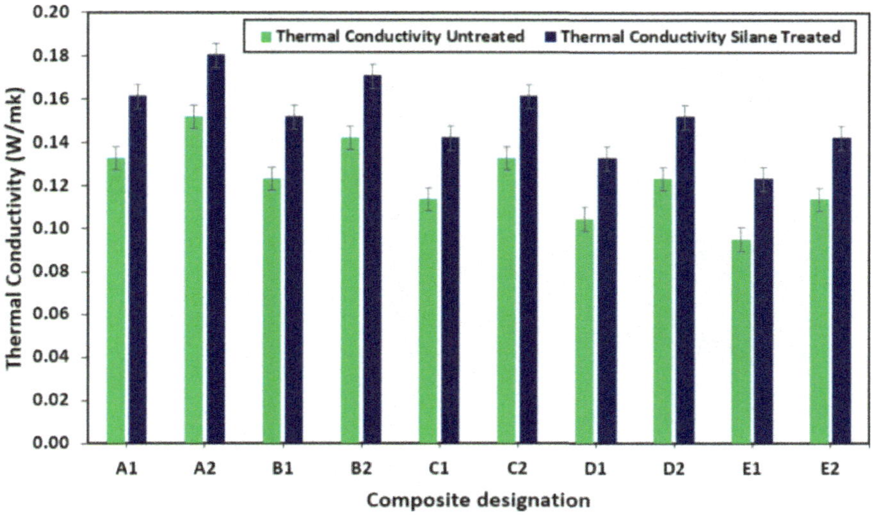

Fig. 4.11 Thermal conductivity of the areca fiber-PET foam reinforced vinyl ester composites exposed to various degradation conditions (Rajeshkannan 2024). Reproduced with permission from Springer Nature

In another study, cenospheres were used as fillers for composites containing untreated and alkali-treated areca husk fibers as reinforcement and PET as matrix. Good bonding was noticed between the matrix and reinforcement, and composites obtained had good thermal insulation and improved hardness. In another study, areca husks were treated with 5% NaOH at 80 °C for 1 h, and fibers obtained were cut into lengths of 3 cm. Later, the fibers (20%) were combined with polyethylene terephthalate (PET) and compounded at 260 °C. Composites were obtained with the desired dimensions suitable for testing. Composites obtained had a microhardness of 27.7 and tensile strength of 57 MPa, flexural strength of 53 MPa, and impact strength of 22.5 kJ/m^2. It was observed that these properties were sufficient for most composite applications and that chemical treatment of the husks was essential to obtain better properties (Nayak 2022).

4.10 PVA as Matrix

Areca husks were cut to length between 10, 20, or 30 mm, which were later combined with PVA in 1:2 ratio for 40 min. The mixture was made into composite samples (Fig. 4.12) having density between 280 and 330 kg/m^3 from which samples having diameters of 30 or 100 mm were cut and used for determining sound absorption properties. It was found that the sound absorption properties varied considerably depending on the composition, density, length of fibers, etc. (Fig. 4.13) (Kalasee

Fig. 4.12 Images of the various samples made for acoustic testing from areca husk fibers combined with PVA (Kalasee 2025). Reproduced with permission through open access publishing

2025). The panels were considered to be suitable for sound absorption in the low to mid frequency region, but presence of PVA made them susceptible to moisture.

In a similar attempt, areca fibers having length of 3 μm and diameter of 800 nm were combined with biochar (80 nm pore size and conductivity of 0.6 to 0.8 ms/cm) prepared from potato peel and cobalt powder and mixed with water soluble PVA (density of 1.19 g/cm^3, and molecular weight of 44.12 g/mol) serving as the matrix. As seen from Fig. 4.14, addition of the biochar and cobalt particles significantly enhanced the electromagnetic shielding (EMI) capabilities of the composites. Further, addition of 0.5% biochar and 1% cobalt particles provided the lowest relative permittivity. Highest tensile strength of 52 MPa and elongation of 108% were also observed when 5% cobalt and 2% biochar were used. Higher content of biochar or cobalt led to agglomeration and hence poor composite properties (Subbayamma 2024). Similarly, composites intended for electromagnetic shielding were produced from polyvinyl alcohol with areca microfibers as reinforcement and rice bran biochar and cobalt particles as additives. Fiber (3 μm in length and 800 nm diameter) content was maintained at 20% with biochar and cobalt content varied between 0.5 to 2 and 0.5 to 5 wt%. As seen from Fig. 4.15, absorption values varied between 14.4 and 46.8 dB, and reflection values between 3.1 and 17 dB were obtained with total shielding effectiveness values between 17.6 and 64 dB. Similarly, relative permittivity was between 6.9 and 0.89 at frequencies between 8 and 20 GHz. These properties were considered to be suitable for developing flexible and high-efficiency EMI shielding materials (Prakash 2024).

In a related study, short areca fibers and carbon quantum dots (2 μm diameter) obtained from corn cobs were added into PVA and made into films suitable for EMI shielding applications. Areca fibers (30% w/w) were supposed to improve mechanical properties, and the carbon dots (0.5 to 4%) would provide the necessary conductivity. Addition of the fibers and the CQDs both led to increase in dielectric constant

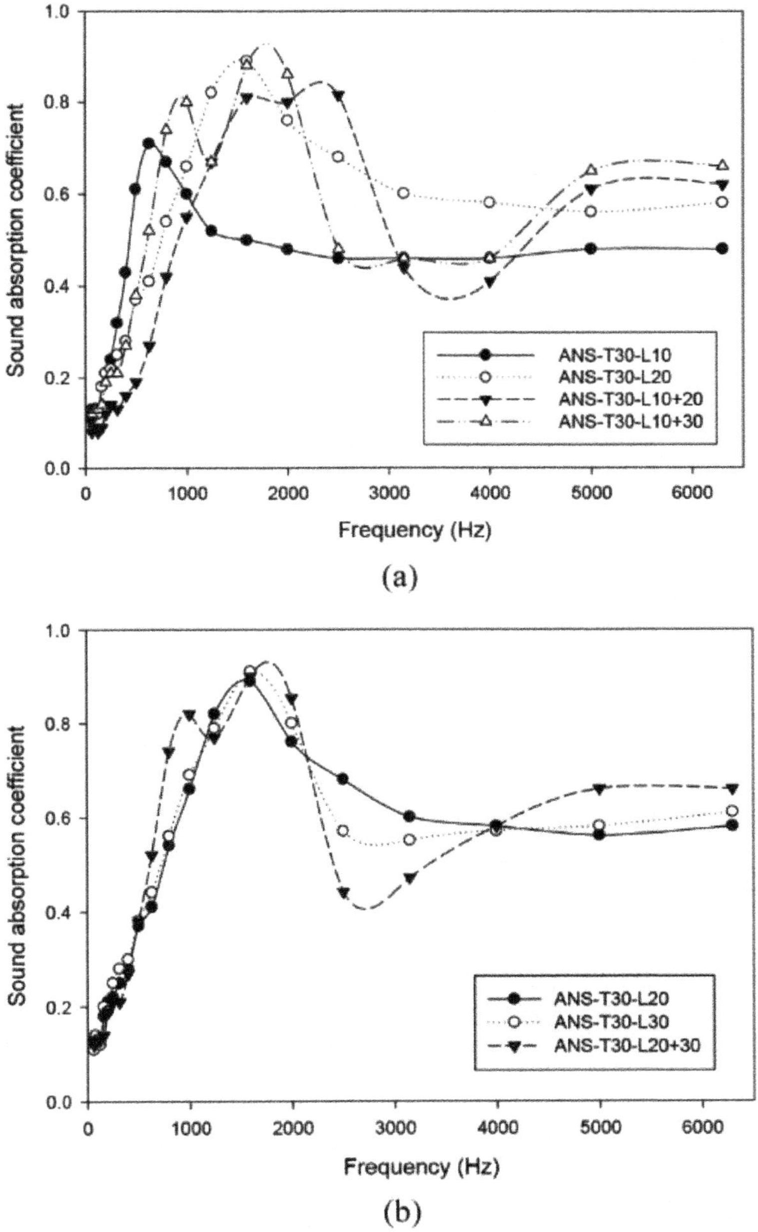

Fig. 4.13 Comparison of sound absorption coefficients of areca fiber with different lengths used as reinforcement for PVA. **a** Fiber lengths of 10 and 20 mm compared with those containing blended fibers. **b** Fiber lengths of 20 and 30 mm compared with those containing blended fibers (Kalasee 2025). Reproduced with permission through open access publishing

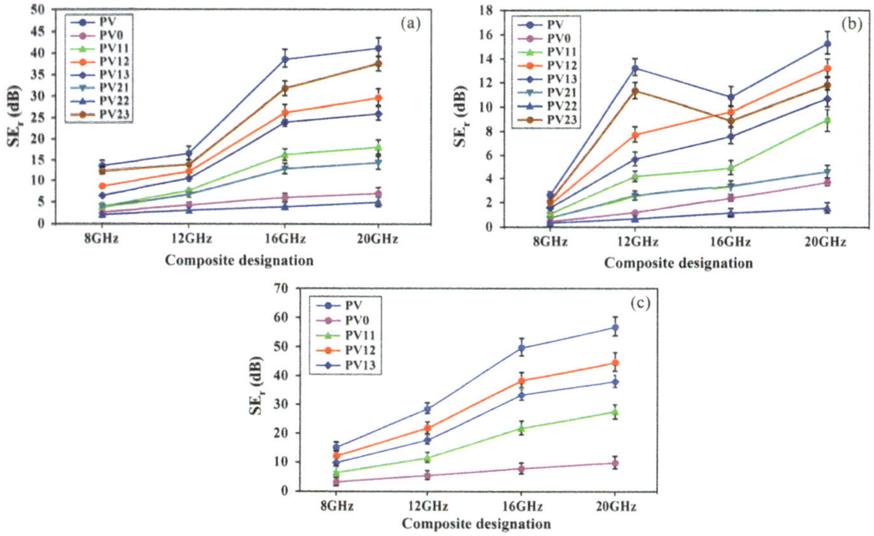

Fig. 4.14 EMI shielding effect for various composites developed from PVA with areca fibers, potato peel biochar, and cobalt in various proportions (Subbayamma 2024). Reproduced with permission from Springer Nature

(Fig. 4.16) and magnetic permittivity. The magnetic permittivity obtained in this study was reported to be the highest ever reported for any composite material. Further, the composites had maximum tensile strength of 58 MPa and maximum shore hardness of 76 and high elongation of 72% (Saravanan 2024). Areca husks were treated with alkali followed by benzyl chloride to obtain fibers having length of 1–5 cm and width of 1 to 10 mm. These fibers were combined with PVA and made into composites through injection molding at 170 °C with fiber content varying from 10, 20, or 30%, and the changes in wear properties were investigated. Wear resistance (specific wear rate) was found to decrease with increasing fiber content or increasing load. A grey-Taguchi method of optimization could be used to develop composites with the desired properties (Nayak 2018).

In a different application, nanocrystals synthesized from areca husk fibers were combined with clay nanoparticles and used as additives for PVA. The nanomaterial-reinforced PVA was made into films for food packaging. About 35% nanocrystals having size of about 232 nm (Fig. 4.17) and zeta potential of −1.5 mV were obtained from the areca fibers, and the nanoclay used had particle size of 50 nm and zeta potential of −0.1 mV. The composites (films) developed had transparency of 0.160, tensile strength of 2.37 MPa, and 240% elongation. With antimicrobial, antioxidant properties and high contact angle of 90°, the films were considered to be suitable for food packaging (Fig. 4.18) (Ramesh 2022).

In a similar approach, cellulose and nanocellulose were extracted from areca husks and used as a reinforcement for poly(vinyl alcohol) films. Addition of 1 to 5% of

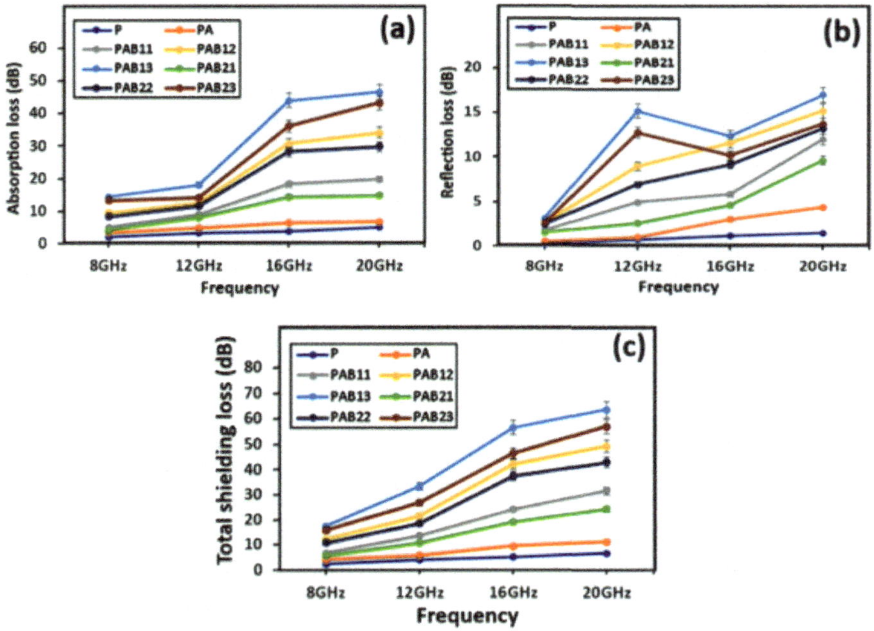

Fig. 4.15 Electromagnetic interference **a** absorption loss, **b** reflection loss, **c** total shielding loss, for the areca fiber reinforced PVA composites containing rice bran biochar and cobalt as additives (Prakash 2024). Reproduced with permission from Springer Nature

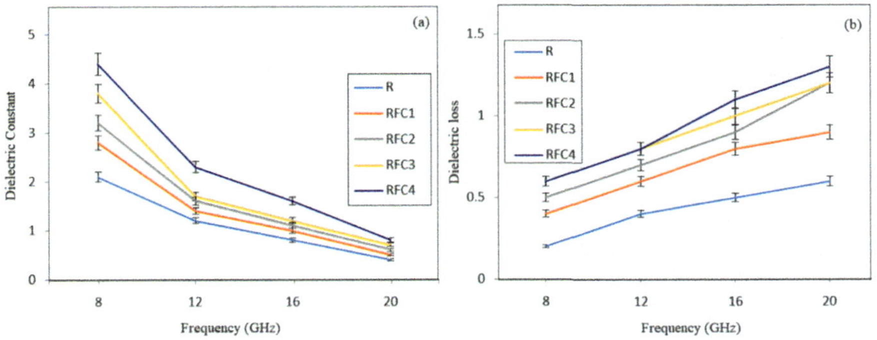

Fig. 4.16 Effect of corn cob carbon quantum dots and areca husk microfiber on EMI shielding effectiveness of flexible PVA thin film at 8–20GHz frequency bands (Saravanan 2024). Reproduced with permission from Springer Nature

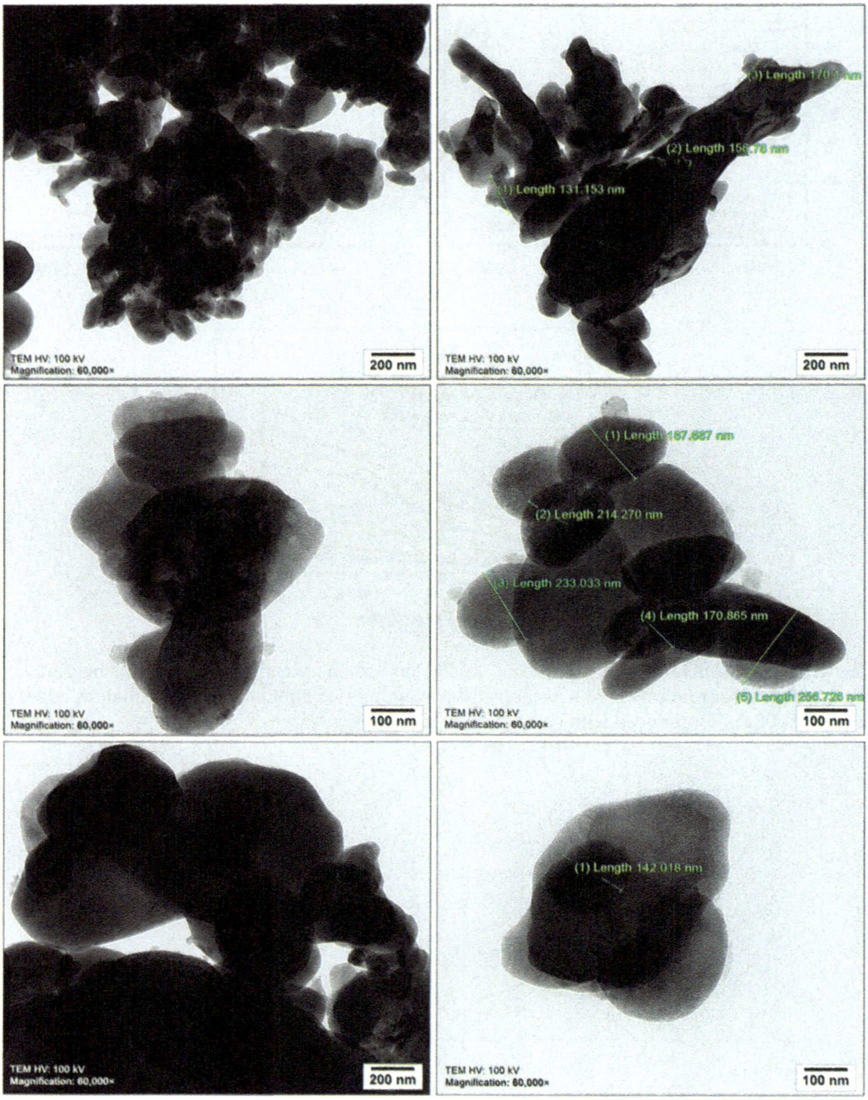

Fig. 4.17 TEM images of arecanut fiber nano crystals (FNC) at scales of 500, 200, 100, and 50 nm (Ramesh 2022). Reproduced with permission from Springer Nature

cellulose or nanocellulose improved the strength but decreased the elongation of the films. Nanocellulose provided higher improvement in strength compared to films reinforced with areca husk cellulose (Fig. 4.19) (Sultana 2020). Thermal resistance of the films was also found to improve, and the films were considered suitable for medical applications.

Fig. 4.18 Digital image of the transparent packaging film made from PVA reinforced with areca husk fiber nanocrystals and nanoclay (Ramesh 2022). Reproduced with permission from Springer Nature

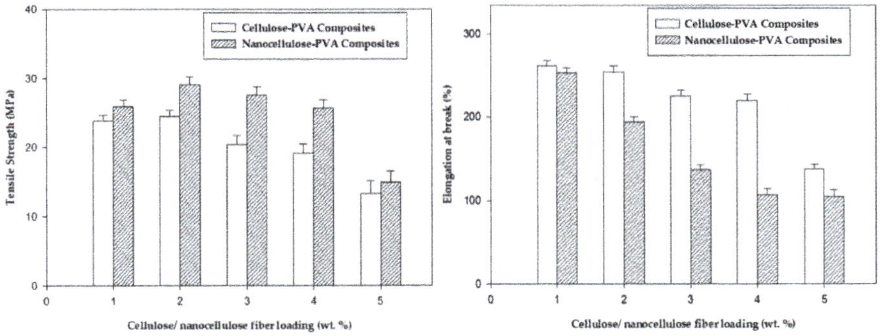

Fig. 4.19 Changes in the strength and elongation of the areca husk cellulose and nanocellulose reinforced PVA films (Sultana 2020). Reproduced with permission through open access publication

4.11 Reinforcement for Poly(Lactic Acid)

PLA is one of the most common biopolymers derived from renewable resources and is relatively easily biodegradable. Hence, several studies have been done to use PLA as a matrix and various fibrous materials as reinforcement, which would produce completely biodegradable composites. In one such attempt, nanocellulose fibers (7–15 nm in diameter) were extracted from areca husks through TEMPO-mediated oxidation and further modified using cis-9-octadecenylamine (oleylamine, OA) to improve compatibility and mechanical properties. Composites were made by adding the nanocellulose fibers into PLA solution, compression molding, and curing with in situ crosslinking. Tensile properties improved with up to 5% addition of the nanofibers, but a substantial improvement in properties could be achieved even when 0.5% of the modified fibers were added as reinforcement. Up to 28% increase in elongation was also possible when 4% modified fibers were used as reinforcement. Reduction in water sorption of the composites was also observed due to the addition of the fibers (Soman 2017). In another effort to develop completely biodegradable composites, areca fiber mats were combined with PLA sheets in multiple layers (Fig. 4.20) and compression molded into composites at 115 °C for 18–30 min at 20-ton pressure. Interestingly, tensile properties including strength, modulus, and toughness were higher for the one ply composites compared to those made using two or three ply probably because of the difficulty in the matrix (PLA) penetrating areca fibers and providing good adhesion (Fig. 4.21). However, flexural strength and modulus were highest for the two-ply composites. The composites developed were considered to be suitable for low-strength structural applications (Kousalyaa 2018).

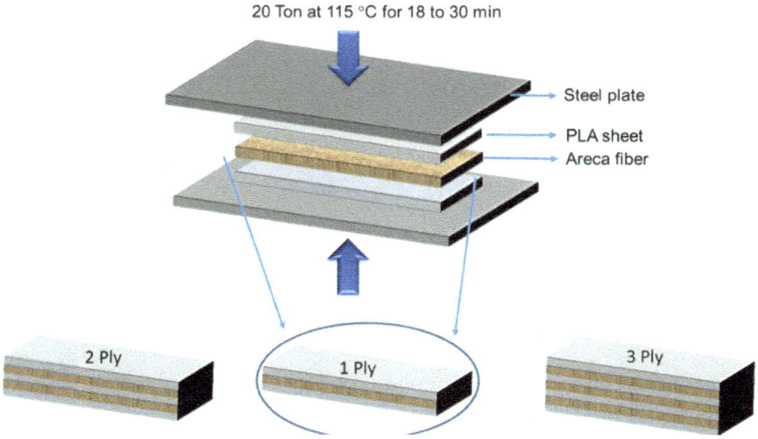

Fig. 4.20 Process of manufacturing the PLA-areca fiber composites of different plies (Kousalyaa 2018). Reproduced with permission from John Wiley and Sons

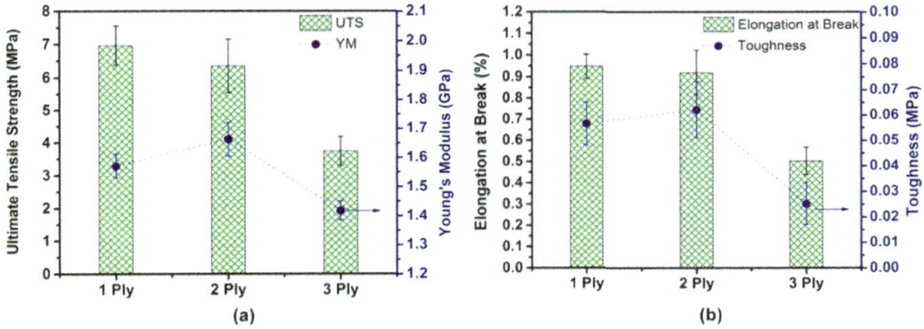

Fig. 4.21 Tensile properties of PLA composites reinforced with multiple layers of areca fibers (Kousalyaa 2018). Reproduced with permission from John Wiley and Sons

4.12 Miscellaneous Matrices

Instead of synthetic polymers, several studies have been done to use natural resins as matrix for areca fiber-based composites. In one such attempt, cellulose nanofibrils were extracted from areca husks and combined with natural rubber latex to form composites. The nanofibrils had aspect ratio of 100–150. The nanofibrils were able to form an extensive network in the matrix, which led to improved mechanical properties. As seen from Fig. 4.22, the percolation threshold of modulus increased with increasing nanocellulose content (Chandra 2022).

Naturally occurring pine resin was used as the matrix, and different ratios of the untreated and alkali-treated areca husk fibers were combined to obtain composites

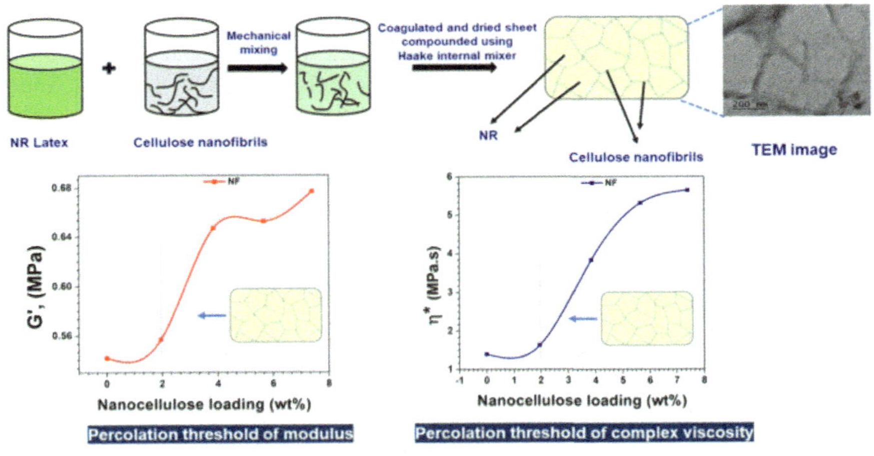

Fig. 4.22 Process of preparing the composites reinforced with arecahusk nanofibrils. TEM image and properties of the composites (Chandra 2022). Reproduced with permission from Springer Nature

Table 4.12 Properties of composite made from areca husk powder and tapioca starch (Nurdin 2020). Reproduced with permission through open access publishing

Areca fiber powder, %	Composite density	Water content, %	Water absorption, %	Modulus of rupture, kg/cm^2	Modulus of elasticity, kg/cm^2	Strong screw hold, kg
90	0.38	9.58	168.02	3.92	552.29	1.19
80	0.45	8.97	121.66	18.20	1837.44	3.68
70	0.57	8.55	71.98	40.61	3475.39	7.19
60	0.71	8.05	51.19	73.31	4761.68	12.88

suitable for various structural applications. Alkali treatment reduced the fiber diameter and increased the interaction between the matrix and reinforcement, leading to better properties. About 33% increase in tensile strength and 24% increase in impact resistance were observed when alkali-treated fibers were used compared to the untreated fibers (Kencanawati 2020).

In another study to develop completely biodegradable composites using areca hut husks, the husks were powdered into particles of about 0.6 mm. The powdered particles were combined with tapioca starch and water. Ratio of the particles in the mixture was varied between 60 and 90%. The mixture was placed in a mold and later compressed at 100 kg/cm^2 and held for 60 min initially and later at a temperature of 120 °C for 60 min (Nurdin 2020). Some of the properties of the composite obtained are given in Table 4.12.

4.13 Mechanical Binding Without Any Matrix

In a unique approach, areca husks were pulped mechanically and combined with alkali-extracted areca husk fibers and made into panels of different densities. Sound absorption coefficient was dependent on the density and the type of panel. For the pure areca sheath panels, the coefficient varied from 0.15 to 0.80 with denser panels providing lower sorption (Fig. 4.23). For the panels containing areca husk fibers, the coefficient varied from 0.19 to as high as 0.9. Thermal conductivity was 0.115 W/mK and up to 0.145 W/mK for the panels made from areca sheath pulp and blend of the pulp and fibers. Panels made only from areca sheath were considered to be suitable for insulation applications (Neelappa 2024).

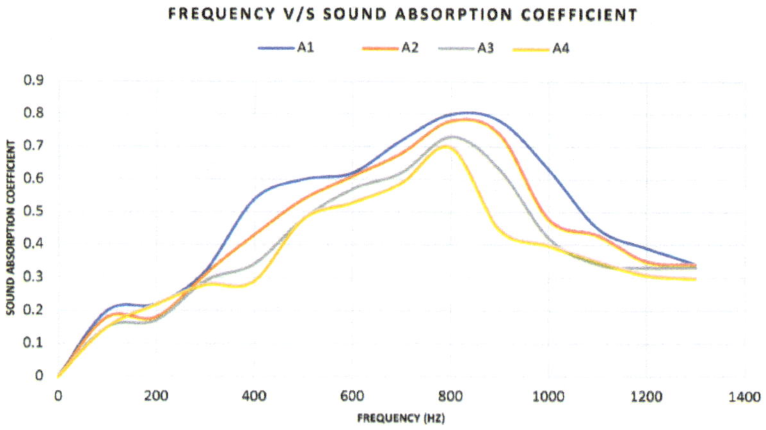

Fig. 4.23 Sound absorption coefficient of panels made from areca sheath pulp of different densities (Neelappa 2023). Reproduced with permission through open access publishing

4.14 Areca Husks as Reinforcement for Concrete

Areca husks are known for their relatively good strength and durability and hence considered as reinforcement for concrete. For this application, the husks were treated in water and manually separated into fibers having average length of 50 mm, diameter of 0.75 mm, and aspect ratio of 66.67. These fibers were combined with cement, aggregates and other constituents with amount of fibers ranging between 1 and 4%. Addition of the husk fibers gradually decreased the slump value by up to 73% as the amount of fiber increased, suggesting decrease in workability due to the addition of the fibers. However, considerable decrease in the density of the concrete was observed. Substituting the cement with fibers created voids, which would decrease the density. In terms of mechanical properties, addition of even 2% fibers increased splitting strength by 18%, and compressive strength was higher by 3–8% (Fig. 4.24). In addition, the fibers resisted cracks due to failure and improved post-cracking performances. Decrease in cost, carbon emission, and low density of the concrete was suggested to be beneficial for reducing the weight of foundations, the cost of building, and promoting green building initiatives (Hasan 2023).

In another study, the feasibility of using areca husk fibers as a stabilizing additive for stone matrix asphalt (SMA) was investigated by Chinnabandar et al. The fibers were combined with the matrix, and the changes in volumetric and Marshall properties were studied in comparison to polymer-modified bitumen. Fibers were able to restrict drain down and required least energy for compaction, suggesting suitability for building applications (Chinnabhandar 2023).

Similar to concrete, alkali-treated areca husk fibers (20–38 mm) have also been used as reinforcement for Compressed Stabilized Earth Blocks (CSEB). The blocks (bulk density between 1810 and 2101 kg/m^3) reinforced with 0.85% fibers showed

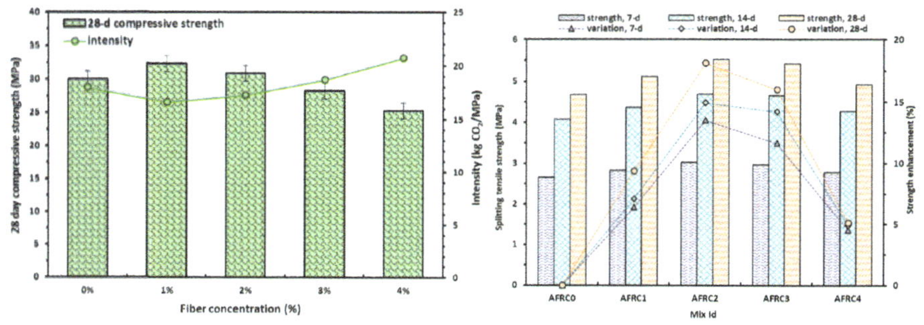

Fig. 4.24 Changes in the compressive strength (left) and splitting strength (right) of concrete reinforced with different ratios of areca husk fibers (Hasan 2023). Reproduced with permission through open access publishing

highest compressive strength of 5.8 MPa after 90 days of curing which satisfied the requirement for class 20 and 30 blocks as per IS specifications. Tensile strength was also higher (0.9 MPa) with 0.85% fibers and curing for 90 days. Reinforcing earth blocks with areca fibers was considered to be technically and economically feasible approach toward sustainable construction (Anik 2024). In another study, adding areca fibers (0 to 3%) into CSEB provided 107–437% higher compressive strength, 208–358% higher split tensile strength and 16–83% higher flexural strength along with up to 100% increase in wet compressive strength. Areca fibers, particularly at 2% were found to provide optimum bonding and stability with reduced voids and fissures. CSEBs manufactured with areca fibers exceeded the standards required for construction and building applications (Paul 2024).

Instead of using directly, husks were pyrolyzed between 300 and 500 °C and added into cement. Pyrolysis led to increased surface area and thermal stability (Fig. 4.25). Addition of the carbon (2%) led to substantial increase in compressive strength of the composites. The strength increased up to 34% when husks carbonized at 500 °C were used. It was suggested that using the biochar in cement would reduce CO_2 emissions by about 22% and help in carbon sequestration efforts (Manjunath 2024).

Addition of up to 2% of areca husk fibers was reported to improve properties of cementitious materials in both experimental and theoretical conditions (Divakar 2020). In another study, areca fibers (50mm length, 1 to 4%) were combined with cement and other additives (Table 4.13) and studied for changes in performance properties (Rashid 2024). Addition of the fibers led to continual decrease in slur slump values from 82 to 22 mm. However compressive strength of the composites showed considerable increase with addition of the fibers specifically as the curing days increased. Highest strength of 21.7 MPa was obtained for 1% fiber containing concrete after 28 days of curing. Fiber content above 3% led to decrease in properties at all curing times (Rashid 2024). Fiber addition also led to smaller crack widths but led to higher sorption of water. Interestingly, eCO_2 emission studies showed (Table 4.14) that addition of areca fibers has advantages compared to conventional concrete.

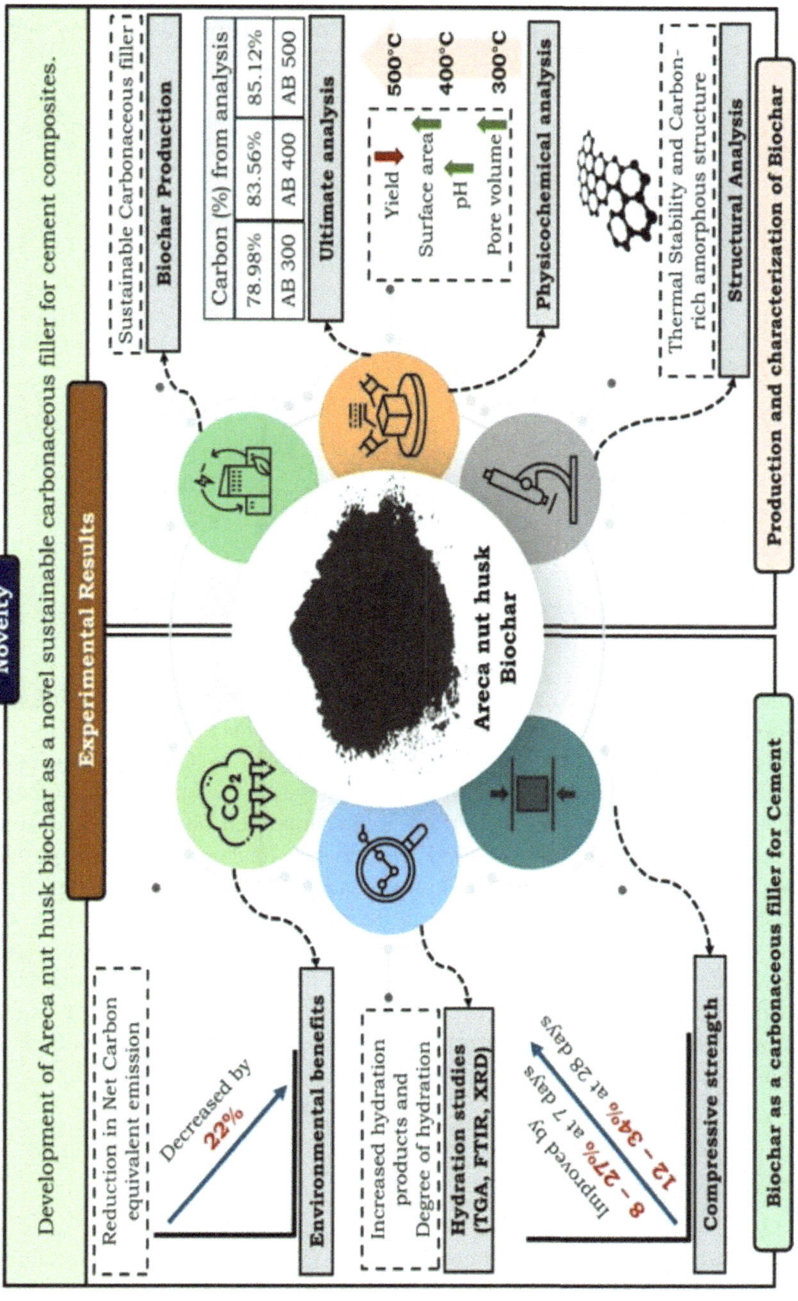

Fig. 4.25 A schematic representation of the process, properties, and performance of areca nut biochar as an additive to concrete (Manjunath 2024). Reproduced with permission from Elsevier

Table 4.13 Composition and percentage of additives used to develop the fiber-reinforced concrete (Rashid 2024). Reproduced with permission through open access publishing

Mix batch	Cement (kg)	Areca fiber (kg)	Fine aggregate (kg)	Coarse aggregate (kg)	Water (liter)	W/C ratio
NFRC0%	13.36	0	22.3	46.8	7.4	0.55
NFRC1%	13.23	0.134	22.3	46.8	7.3	0.55
NFRC2%	13.09	0.267	22.3	46.8	7.2	0.55
NFRC3%	12.96	0.402	22.3	46.8	7.1	0.55
NFRC4%	12.82	0.536	22.3	46.8	7.0	0.55

Table 4.14 Energy, eCO_2 emission during production and transportation for cement (Rashid 2024). Reproduced with permission through open access publishing

Mix batch	eCO_2 emission (kg CO_2/ kg concrete)			Percentage of CO_2 emission	
	Production	Transportation	Total	Production (%)	Transportation (%)
NFRC0%	255.77	94.73	350.51	72.97	27.03
NFRC1%	253.35	94.68	348.03	72.80	27.20
NFRC2%	250.74	94.62	345.36	72.60	27.40
NFRC3%	248.32	94.57	342.88	72.42	27.58
NFRC4%	245.70	94.51	340.21	72.22	27.78

Similar to using areca husks directly and also fibers from the husks as reinforcement of concrete, asphalt, etc., the feasibility of using areca husk fibers as a stabilizing additive for stone matrix asphalt (SMA) was investigated by Chinnabandar et al. Husk fibers were combined with the matrix, and changes in volumetric and Marshall properties were studied in comparison to polymer-modified bitumen. Fibers were able to restrict drain down and required least energy for compaction, suggesting suitability for building applications (Chinnabhandar 2023). Fibers extracted from areca husks were combined with cement in 0, 0.25, 0.5, 0.75, and 1% (w/w), and the effect of fiber addition on the workability, mechanical strength, shrinkage, and porosity was determined in comparison to adding jute or coir into mortar. Addition of areca fibers up to 0.5% increased mechanical strength but decreased at higher levels. Considerable reduction in shrinkage and increase in porosity were observed with 0.5% fiber addition, giving the most optimum properties (Miah 2023).

4.15 Conclusions

Areca husks are lignocellulosic and could be preferred for developing into natural cellulose fibers and composites. Considerable work has been done on utilizing areca husk, and the fibers extracted from the husks as reinforcement for composites.

Although epoxy has been the predominant matrix, new biodegradable matrices such as poly(lactic acid) or protein-based binders have been used to develop partially or fully biodegradable composites. Several natural and synthetic additives have also been incorporated to enhance the properties of the areca husk-based composites. As a unique approach, areca husks have also been successfully used to improve the performance properties of concrete. Despite such wide applications, it is necessary for further studies to use physical and/or chemical methods to improve the properties of areca husks and ensure wider applications.

Chapter 5
Areca Leaf Sheath

5.1 Introduction

Areca leaf sheaths are the integral part where the leaf attaches to the stem. Sheaths are probably the highest component by weight and most widely used byproduct generated from the tree. Each areca nut tree sheds 5–6 leaves per year, and about 2.33 lakh tons of the sheath are generated annually. The sheaths are mainly composed of α-cellulose (66%), lignin (19%), and hemicelluloses (7%), considerably different than the other parts of the plant. However, some researchers have reported lignin content in the sheaths to be as high as 39% (Banagar 2018). Sheaths are obtained in different sizes with average length of 81 cm, width of 28 cm, and thickness of about 4 mm. In terms of mechanical properties, raw areca sheaths were reported to have a flexural strength of 42 N and tensile strength of 16 MPa, and areca sheaths were also reported to have good electrical conductivity. As with the other parts, areca sheaths also show considerable variations in the composition, structure, and properties of the sheaths depending on the growing region, plant variety, and age, etc.

As seen from Table 5.1, crude fiber content varied from 28.7 to 33.7%, whereas lignin content was similar. Areca leaf sheaths were reported to contain approximately 10% more fibers, 7–8% more iron, and 4–5% more magnesium than common biomass such as rice straw, which is used as animal feed (Prasannakumaran 2024). The tensile and flexural strength of the sheaths also vary considerably when subject to chemical treatments, particularly after acrylation or treating with alkali (Banagar 2018). Studies have been done to understand the mechanical properties of areca sheaths before and after exposure to sunlight and to chemicals. Considerable changes in thickness, density, surface morphology, and mechanical properties were observed depending on whether the sheath was sun-dried, oven-dried, or treated with alkali. Age of the areca plants from which the sheaths were collected also influenced the tensile and flexural properties (Table 5.2) (Banagar 2021). Density of the sheaths varied from 0.536 to 0.584 g/mm^3 depending on the age. Oven-dried sheaths provided better strength than sun-dried samples (Table 5.2).

N. Reddy and P. Aramwit, *Areca catechu and its byproducts*,
https://doi.org/10.1007/978-3-032-22134-6_5

Table 5.1 Tensile strength of the areca sheath collected from plants of different ages after sun drying (SD) or oven drying (OD) and without treating or treating with 5 and 10% alkali (NaOH) (Banagar 2021)

Age, years	UT-SD	UT-OD	5% SD	5% OD	10% SD	10% OD
3	19.33	12.93	14.90	14.40	15.22	17.50
4	15.67	15.78	25.48	30.27	18.56	18.04
5	18.45	19.58	15.77	24.69	20.97	19.00
6	18.87	12.31	20.19	18.60	26.75	28.61
7	17.31	25.69	24.40	28.80	33.58	35.83

5.2 Natural Cellulose Fibers from Areca Sheaths

Several studies have shown the possibility of extracting natural cellulose fibers from areca sheaths. In a simple process (Fig. 5.1), natural cellulose fibers resulting in 57% cellulose, 22% hemicellulose, and 6.3% lignin were obtained from areca sheaths. Properties of the fibers obtained from areca nut sheaths are compared to other common fibers in Table 5.3 (Das 2020). Fibers obtained had diffraction peaks similar to those of cellulose seen in common lignocellulosic fibers with % crystallinity of 56.5% and microfibrillar angle of 35.2°. Although fibers from areca sheaths had most properties similar to lignocellulosic fibers from other biomass sources, their shorter length and coarser diameter make it difficult for the fibers to be processed as typical textile fibers (Raj 2020). However, the fibers could be useful to develop composites, for manufacture of paper and pulp, and for thermal and noise insulation applications. In another study, fibers extracted from areca sheaths through water retting for 15 days had cellulose content of 66% and lignin content of 20% (Poddar 2016) (Table 5.4), and it was suggested that the areca sheath fibers had higher cellulose content than fibers obtained from coconut leaf sheaths. In another study, although the process used for extraction was not reported, fibers with cellulose content of 65% and lignin content of 18.6% were obtained. Length of the fibers was between 800 and 1400 mm, diameter was between 400 and 700 μm, and linear density was 214 tex. Further, X-ray diffraction studies showed that the fibers had a crystallinity of 81%. Tensile strength of the fibers was reported to be 60 MPa, and the fibers had good thermal stability with residual mass of 33% after heating to 500 °C. The fibers were suggested to be suitable for reinforcement in composites used for automobiles, interior, and other weight-bearing structures (Suresh 2021). Properties of natural cellulose fibers obtained from areca sheaths could be further modified using various chemicals. For instance, tensile strength was found to increase particularly after treating with benzoyl chloride, and the modulus of the fibers was found to be highest when treated with acrylic acid. Considerable and distinct changes were also observed in the surface morphology of the fibers (Fig. 5.2). A large number of perforations and holes and general increase in surface roughness could be observed after

Table 5.2 Characteristics of areca sheath industry wastes samples collected from various locations (A-D) (Prasannakumaran 2024)

Samples	Moisture content %	Crude protein %	Crude fiber %	Ash content %	Ether extract %	Lignin %	NDF %	ADF %	TPC %
A	10.22 ± 0.2^{a}	4.37 ± 0.2^{a}	32.12 ± 0.9^{abc}	2.49 ± 1.3^{b}	3.96 ± 0.6^{a}	7.8 ± 0.13^{a}	71 ± 1.64^{ab}	43 ± 2.76^{b}	1.43 ± 0.30^{c}
B	10 ± 0.4^{a}	4.8 ± 0.43^{a}	28.65 ± 0.5^{c}	3.79 ± 0.1^{a}	3.33 ± 0.2^{a}	8.12 ± 0.25^{a}	69.78 ± 1.27^{b}	44.07 ± 1.76^{ab}	2.60 ± 0.50^{ab}
C	9.78 ± 0.3^{a}	4.4 ± 0.03^{a}	29.89 ± 0.3^{bc}	3.35 ± 0.4^{ab}	3.62 ± 0.4^{a}	8.01 ± 0.11^{a}	71.64 ± 0.10^{ab}	49.3 ± 2.4^{a}	2.72 ± 0.50^{a}
D	10 ± 0.39^{a}	4.36 ± 0.11^{a}	33.71 ± 1.4^{ab}	3.15 ± 0.6^{ab}	3.96 ± 0.6^{a}	7.9 ± 0.5^{a}	73.77 ± 1.85^{ab}	44.84 ± 1.30^{ab}	1.58 ± 0.13^{bc}

Mean ± SE ($n = 2$) values that do not share the same superscript alphabet (a–g) in a column are significantly different ($p < 0.05$). A—Paramakudi, B—Palakkad, C—Coimbatore, D—Mangalore

Reproduced with permission through open access publishing

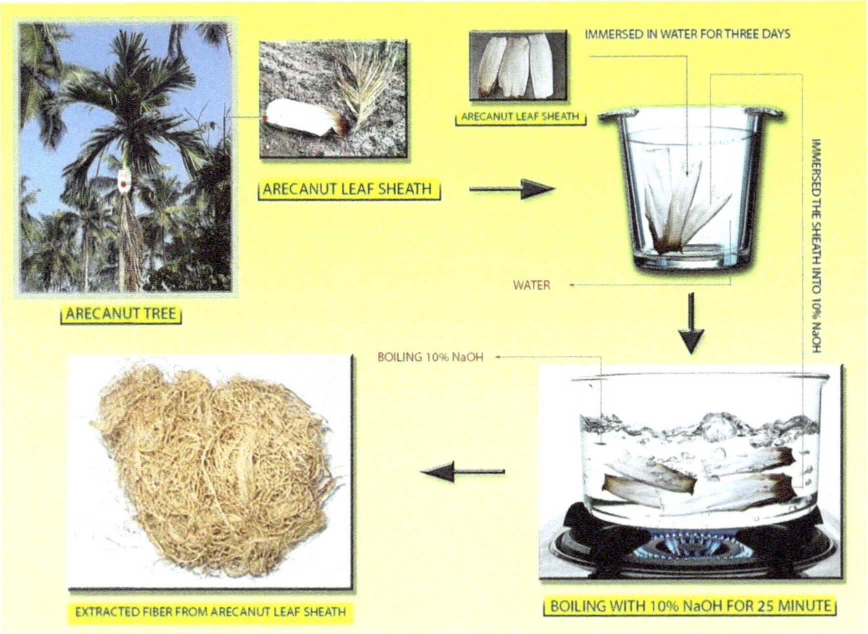

Fig. 5.1 Process used for the extraction of natural cellulose fibers from areca sheaths (Das 2020). Reproduced with permission from Taylor and Francis

the chemical treatment (Fig. 5.2) (Nayak 2019). Natural cellulose fibers with properties similar to or better than those of chemically treated fibers have been obtained from areca sheaths using enzymes (Sunny 2022). It has been demonstrated that the properties of the natural cellulose fibers obtained from the areca leaf sheaths depend on the conditions used for extraction (Table 5.5) (Loganathan 2020).

5.3 Areca Sheaths as a Source for Nanocellulose

Few studies have explored the potential to extract cellulose from areca sheaths in nanoform. In one such attempt, areca sheaths having about 72% cellulose were subject to acid hydrolysis using 64% sulfuric acid for the production of cellulose nanowhiskers. The nanowhiskers obtained had a crystallinity index between 82 and 86%, and crystallite size was between 5.69 and 6.36 nm (Elanthikkal 2021), the degree of polymerization was 19 to 78, and molecular weight was 3055 to 12,631, which varied considerably with changes in heating temperature and time. As seen from Fig. 5.3, nanowhiskers with diameter as low as 34 nm were obtained. Particle size analysis showed that the average particle size was 540 nm. Treating areca sheaths with 10% NaOH for 48 h resulted in fibrous pulp, which was compressed into

Table 5.3 Comparison of the properties of natural cellulose fibers obtained from areca sheaths (Das 2020)

Fiber property	Cotton	Jute	Flax	Hemp	Ramie	Coir	Sisal	Areca sheath
Diameter, μm	16–21	30–140	40–620	16–50	40–60	100–450	100–300	54.8 ± 60
Moisture content, %	7.0–8.5	12.6	10.0	10.9	6.8	10.8	10–22	11.9
Cellulose content, %	94	61–71.5	71	70.2–74.4	68.6–76	43	65.8	56.8
Hemicellulose, %	–	13.6–20.4	18.6–20.6	17.9–22.4	13.1–15.0	0.3	12.0	22.4
Lignin, %	–	12–13	2.2	3.7–5.7	0.6–1	4.5	9.9	6.3
Crystallinity, %	68	55.0	72.6	87.87	70	–	17.7	56.5
Orientation angle	20–34	8.0	10.0	6.2	7.5	45.0	20.0	35.2
Density, g/cm^3	1.5–1.6	1.3	1.5	1.5	1.5	1.2	1.3–1.5	1.44 ± 0.01
Tensile strength, MPa	287–597	393–773	345–1035	550–1110	400–938	175	511–635	401 ± 2.34
Young's modulus, GPa	55–126	26.5	27.6	30–60	44–128	4.0–6.0	9.4–22.0	8.3 ± 2.34
Elongation, %	7–8.0	1.5–1.8	2.7–3.2	1.6	2.0–3.8	30.0	2.0–2.5	7.8 ± 2.57

Reproduced with permission from Taylor and Francis

Table 5.4 Composition of areca leaf sheath (ALS), coconut leaf sheath (CLS), and coconut stem fibers (CSF) (Poddar 2016)

Name	%		
	ALS	CLS	CSF
Aqueous extract	0.72	0.53	5.52
Fatty and waxy matters	5.06	2.86	5.91
Pectic matters	1.15	0.80	1.75
Lignin	19.59	27.97	22.54
α-cellulose	66.08	59.39	46.26
Hemicellulose	7.40	8.45	18.02
Total	100	100	100

sheets of various densities (400–700 kg/m^3). Excellent thermal and noise insulation was obtained depending on the density of the compressed sheet. Similarly, thermal conductivity ranging from 0.033 to 0.115 w/mK were obtained, suggesting suitability of the panels for acoustic and insulation applications (Neelappa 2023). Cellulose nanofibers were also obtained by treating the sheaths with alkali, bleaching with 15% hydrogen peroxide, and subjecting them to steam explosion. Later, the treated material was acid hydrolyzed using oxalic acid steam explosion at 120 °C for 4 h to form cellulose nanofibers. These nanofibers were combined with dried ZnO nanoflakes and stearic acid to impart hydrophobicity and surface roughness and were cast into paper. Nanofibers formed had an average diameter of 20 nm, and the ZnO particles were 26 nm. Adding ZnO and stearic acid substantially increased the hydrophobic properties, with contact angle increasing from 34° to as high as 149°. The tensile strength of 22.6 MPa for the areca-based paper was higher than that of commercial filter paper with similar biodegradability, suggesting suitability for packaging applications (Poulose 2024).

5.4 Pulp and Paper

The possibility of using areca sheath for the production of paper boards was studied by Raghupathy et al. (2002). Areca sheaths were soaked in water and mechanically processed to form a pulp. Areca sheaths were made into paper using the pulp and also by combining with waste paper in different ratios. The papers obtained were made into boards and tested for their performance properties according to Indian Standards and were found to satisfy the requirements. In a different approach, areca sheaths were subject to oxalic acid treatment and later steam exploded to form cellulose nanofibers for potential use as nanopapers for packaging applications. Nanocellulose obtained was combined with stearic acid and 2% ZnO nanoflakes and made into paper. The paper formed was considerably hydrophobic with contact angle of 134 °C due to the hydrogen bonding between cellulose and Zinc Oxide nanoflakes. Further, esterification reaction between carboxyl groups in stearic acid and ZnO leads to the

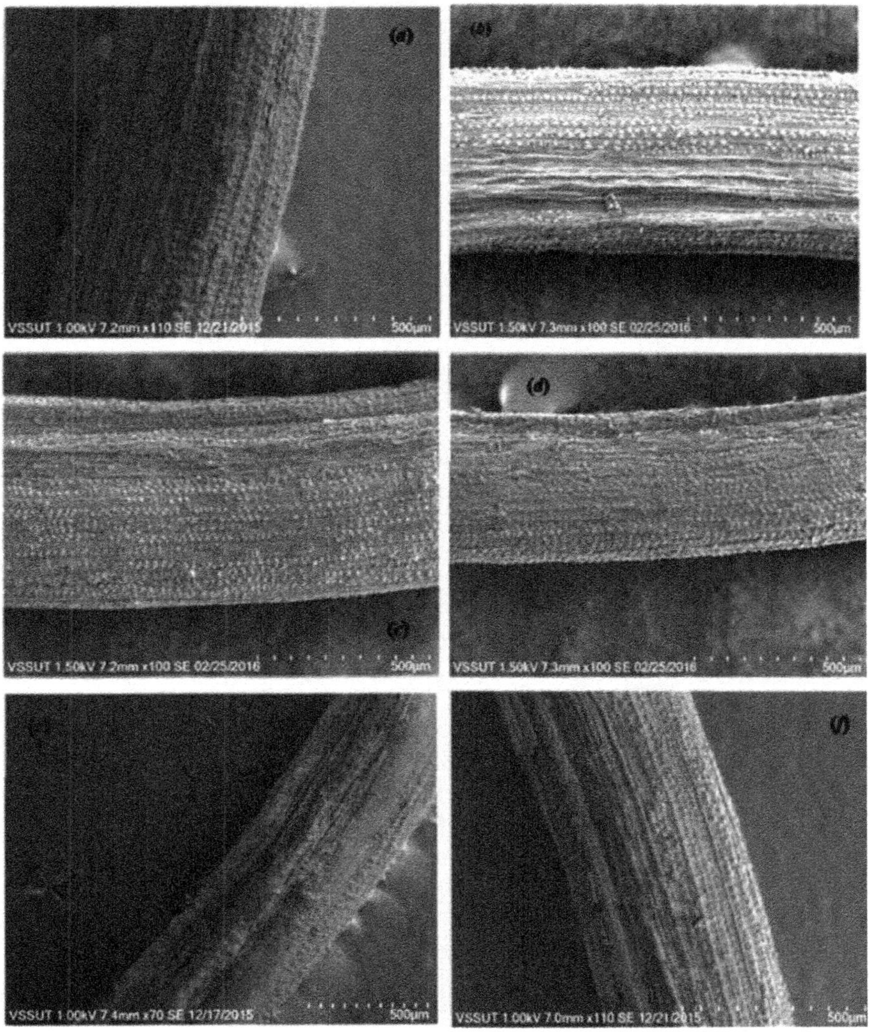

Fig. 5.2 SEM images of **a** untreated and areca sheath fibers subject to **b** alkaline, **c** sodium chlorite, **d** permanganate, **e** acrylic acid, and **f** benzoylation treatments (Nayak 2019). Reproduced with permission from Taylor and Francis

formation of an insoluble stearate layer, adding to the hydrophobicity of the paper. The paper obtained had a considerably high strength of 22.4 MPa, excellent UV blocking ability, and hence preferable for food packaging applications (Poulouse 2024).

Table 5.5 Comparison of the composition of areca fibers obtained from the leaf sheaths using different approaches (Loganathan 2020).

Fiber	Cellulose, %	Hemicellulose, %	Lignin, %	Wax, %	Moisture, %
Husk fiber	53.20	32.98	7.20	0.64	
Areca leaf stalk fibers	57.49	18.34	7.26	0.71	9.35
5% NaOH treated areca leaf stalk fibers	68.54	6.13	5.87	0.32	3.62
10% NaOH treated areca leaf stalk fibers	65.90	–	5.31	0.28	2.84
15% NaOH treated areca leaf stalk fibers	63.45	–	4.54	0.20	1.15
Areca sheath/frond fiber	66.08	7.40	19.59	5.06	0.72
Areca frond fiber	35.90	26.60	16.60	11.70	9.20
Untreated raw husk fiber	43.49	17.71	26.85	1.69	–
Untreated ripe husk fiber	43.99	12.07	26.88	3.13	–
Untreated matured husk fiber	50.21	13.16	29.34	3.48	–
5% NaOH treated raw husk fiber	49.26	8.49	22.17	4.54	–
5% NaOH treated ripe husk fiber	52.09	12.04	22.34	5.90	
5% NaOH treated matured husk fiber	50.51	12.37	26.1	2.53	–
Areca	57.35–58.21	13.0–15.42	23.2–24.2	0.12	7.32
Coconut	32.0–43.80	0.15–20.00	40–45.0	–	8.0
Palm	60.0–65.0	–	21.0–29.0	–	

Reproduced with permission from Springer Nature

5.5 Areca Leaf and Sheath as Sources for Lignin

Areca leaf and sheaths have been used as a source for lignin. ALS were subject to kraft pulping, and lignin was separated from the black liquor using 20% phosphoric acid. A high yield of 12.7% with purity of 62.7% was obtained with the lignin showing potential use as adhesive or filler for biobased products (Ridho 2023). In another study, it was shown that using HCl for lignin precipitation provided higher yield (38–40%) compared to phosphoric acid (14–15%). Lignin obtained had similar composition and structure compared to that obtained from other materials and processes (Ridho 2024). Further, nanolignin having particle size of 227 nm and zeta potential of −18.57 mV could be extracted from areca leaf sheath using high shear homogenization at 13,000–16,000 rpm for 1–4 h. Obtained lignin was used as an additive to urea

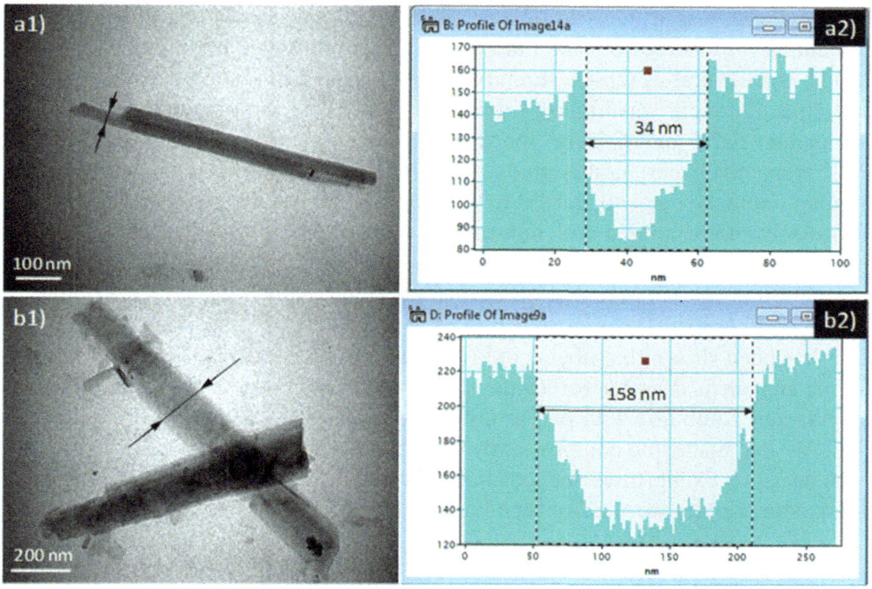

Fig. 5.3 SEM images of the cellulose nanowhiskers obtained from areca leaf sheaths at two different hydrolysis time and temperatures (Elanthikkal 2021). Reproduced with permission from Taylor and Francis

formaldehyde resin and when used as adhesive provided higher tensile shear strength (1.07 MPa) compared to other resins used for adhesion of plywoods and also had lower formaldehyde emissions (Ridho 2024b).

5.6 Composites Using Areca Sheaths

Areca sheaths have been extensively studied for use in composites both in their original form and also after extracting natural cellulose fibers. In a study on developing composites from areca sheaths, the sheaths were collected and treated with sodium hydroxide (5%) at 80 °C for 1 h. Fibers obtained were washed and dried at 60 °C for 24 h and later cut into 3 mm pieces. The treated fibers were combined with PET matrix in a micro compounder at 260 °C at 100 rpm for 5 min and made into composites by varying the proportion of fibers (10 or 20%) (Nayak 2020). Fibers obtained from areca sheaths were also combined with moringa fibers in 50/50 ratio and made into composites. Areca sheath fibers had strength of 66 MPa before treatment but increased to 92 MPa after treating with alkali. When combined with PET and made into composites, the tensile strength of the composites was 57 MPa compared to 50 MPa for pure PET, whereas the areca sheath fiber/moringa fiber composite had marginally higher tensile strength of 61 MPa. Reinforcing PET with the fibers marginally increased the flexural strength from 50 to 53 MPa, but the composites containing a blend of both the fibers had considerably higher strength of 78 MPa. Unlike the flexural or tensile strength where marginal improvement was observed, the impact strength was highest for composites having the blend fibers (35 kJ/m^2) compared to 22.5 and 17 kJ/m^2 for the areca sheath containing composites and neat PET, respectively. In another study, fibers were extracted from areca sheath after treating with water and alkali (8% w/w, 1:15 ratio for 20 min). The fibers were made into sheets and combined with unsaturated polyester resin, accelerator, and catalyst and placed in mold and compressed into sheets at 32 °C for 60 min (Fig. 5.4). Properties of the composites were dependent on the amount of fibers, as seen from Table 5.6, with 25% fiber loaded composites providing the best properties (Das 2021).

It has been shown that fibers could be manually extracted from areca sheaths by immersing them in water for 15 days. The fibers obtained were cut into 2–3 mm lengths and combined with polypropylene granules (50–60 μm) in different ratios (5–25%) and made into composites by compression molding at 190° for 5 min under 5 bar pressure. Composites obtained had strength ranging from 23.1 to 28.7 MPa. The composite samples were treated at 65 and 45 °C in 4 and 2-h cycles for a total time of

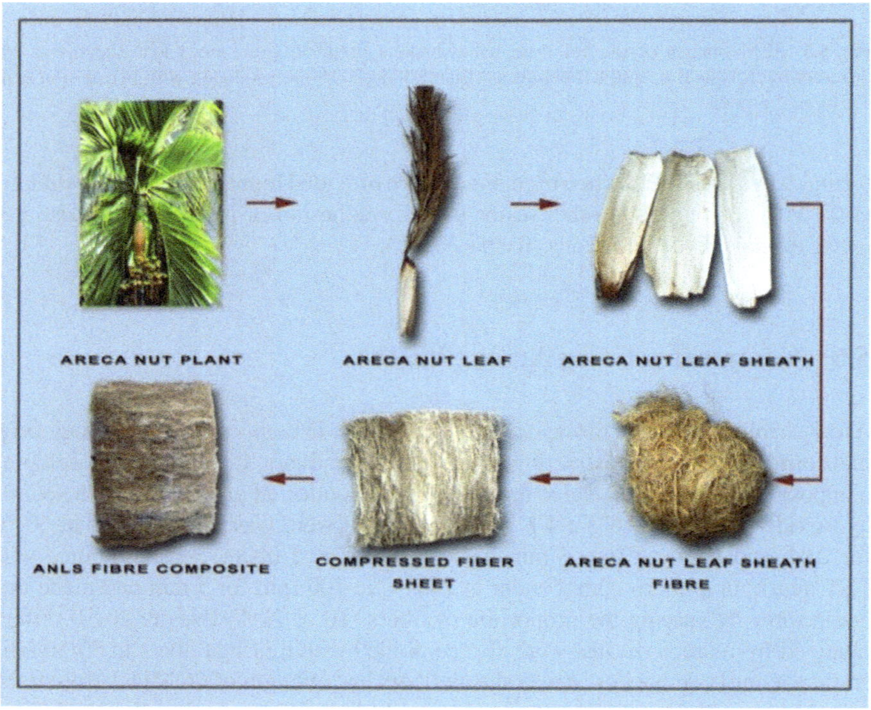

Fig. 5.4 Process used to prepare the composites from areca leaf sheath fibers and polyester resin (Das 2021). Reproduced with permission from Taylor and Francis

Table 5.6 Comparison of the properties of the polyester resin composites reinforced with three different levels of areca sheath fibers (Das 2021)

Properties	15%	20%	25%
Tensile strength, MPa	39.6 ± 5.27	52.22 ± 6.86	68.52 ± 6.34
Elongation at break, %	5.5 ± 0.7	5.00 ± 0.51	4.42 ± 0.59
Tensile modulus, GPa	1.73 ± 0.14	2.24 ± 0.50	2.82 ± 0.54
Flexural strength, MPa	51.5 ± 4.65	63.08 ± 7.29	73.32 ± 7.66
Flexural modulus, GPa	2.23 ± 0.30	2.92 ± 0.61	3.78 ± 0.63
Interlaminar shear strength, MPa	6.25 ± 0.52	7.24 ± 0.78	8.84 ± 0.77
Void content, %	6.4	7.2	5.3

Reproduced with permission from Taylor and Francis

100 h to understand the ability of the samples to withstand weathering. It was found that the composites had good weathering capacity with maximum decrease in strength of 22% (Poddar 2018). In another study, fibers were extracted from areca sheaths after immersing in water for 7–10 days and later combined with epoxy resins and hardener and made into composites using the hand lay-up technique at temperatures between 140 and 150 °C and pressure between 100 and 100 bar. Composites developed were able to withstand rotor dynamic frequency and had a damping factor higher than 1, suggesting suitability as a porous sound absorber (Chethan 2018). Similarly, reinforcing epoxy with fibers (30%, 30 mm length) obtained by soaking areca sheaths in water for 4 to 5 days provided composites with tensile strength of 16 MPa and flexural strength of 3345 MPa (Chethan 2016). However, the extent of improvement in the properties of the composites was dependent on the fiber fraction. Adding 12 to 48% of short areca sheath fibers into epoxy provided considerably enhanced properties. For instance, increase in void fraction from 0.26 to 6.14%, hardness of 72.56 HRB, Young's modulus of 1.98 GPa and impact strength of 4.7 J were obtained but addition of fibers led to decrease in flexural strength and flexural modulus compared to neat epoxy (Suresh 2021). These composites were studied for their tribological behavior and they were found to have minimal wear rate at high sliding velocity of 1.2 m/s. Higher coefficient of friction (COF) (Fig. 5.5) and specific wear rate were obtained for the areca sheath fiber containing composites (Suresh 2021b). Further, developing hybrid composites through the addition of basalt fibers and fillers like clay and Al_2O_3 provided higher performance compared to using only the areca sheath fibers (Suresh 2021c). In another study, combining areca sheath particles (10% by weight) into epoxy matrix increased tensile strength and modulus, impact strength, and hardness, making the composites suitable for partition panels, ceiling panels, and other applications (Varghese 2025). Alternatively, fibers obtained from areca sheaths were made into a plain woven fabric and used as reinforcement for epoxy composites. Fibers used had length between 820 and 1400 mm and diameter between 400 and 700 μm, and linear density was 0.90 g/cm^3. Tensile strength, modulus and elongation at break were 59.62 ± 2.20 MPa, 1.24 ± 0.12 GPa, and 3.02 ± 0.36%, respectively. These fibers were woven into a fabric and layers of the fabrics were placed in a mold and

the matrix epoxy (LY556, density: 1.15 g/cm^3) and the hardener HY-951 were added and later compression molded at 40 atm pressure for 24 h to form the composites. Fiber content in the resin was varied from 0 to 48% resulting in composites with density between 1.089 and 1.147 g/cm^3 and void fraction between 0.26 and 4.9% (Subramanyam 2024). Some of the mechanical and performance properties of the composites are given in Table 5.7. The composites were found to have properties better than similar composites made using other biomass. Samples with 36% fiber content showed lower wear resistance and coefficient of friction.

Soaking areca sheaths in water for 10 days and subsequent manual extraction resulted in fibers having length of 15–20 cm. These fibers were composed of 35–65% hemicellulose and 13–24% lignin and used as reinforcement for composites. Obtained fibers were chemically treated to change the surface and improve interfacial

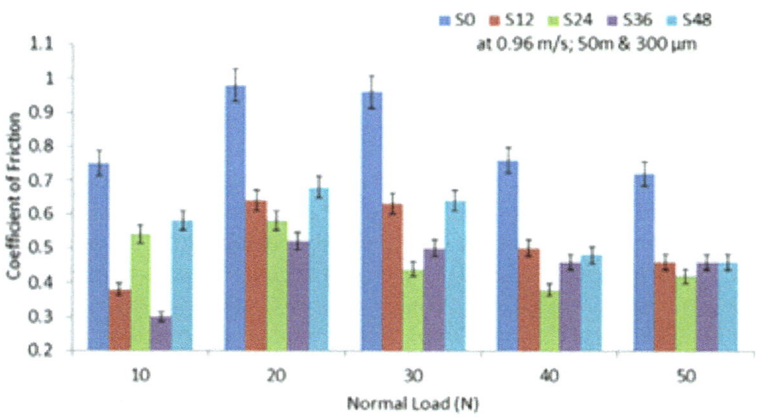

Fig. 5.5 Changes in the normal load Vs coefficient of friction (COF) for epoxy composites reinforced with 12–48% of short areca sheath fibers at three different sliding rates (Suresh 2021b). Reproduced with permission from Elsevier

Table 5.7 Comparison of the tensile and performance properties of epoxy composites reinforced with different ratios of areca fiber woven fabric as reinforcement (Subramanyam, 2024)

Specimen	Hardness (HRB)	Tensile strength (MPa)	Tensile modulus (GPa)	Flexural strength (MPa)	Flexural modulus (GPa)	Impact strength IS (J/m)	Interlaminar shear strength ILSS (MPa)
M0	40.23	43	0.15	45.16	2.25	52.06	1.02
M12	68.26	68.26	0.68	26.84	0.524	28.02	2.84
M24	74.86	88.68	2.88	34.52	0.682	42.64	3.86
M36	78.62	96.82	3.24	50.96	1.18	56.24	4.26
M48	84.26	104.46	4.22	52.84	2.86	64.26	4.68

Reproduced with permission through open access publishing

interactions. For the chemical treatment, fibers were soaked in 5% NaOH solution for 1 h at 80 °C. Later, the fibers were washed to remove the alkali and further treated with benzoyl chloride for 15 min to increase the hydrophobicity and interaction between the fibers and matrix. Composites were developed by mixing up to 40% fibers (cut into 3 mm) with polyvinyl alcohol at 170 °C for 5 min. This mixture was passed through an injection mold and square-shaped composites were developed (Nayak 2020). Erosion rate of pure PVA and the areca fiber reinforced composites were studied at different impact angles and impact velocities. Unreinforced PVA had highest erosion rate of 30% whereas minimum erosion was obtained for the 10% fiber containing composites. Fiber-reinforced PVA had considerably lower impact even when tested at an impact angle of 45° and it was found that 27% reinforcement of fibers provided better mechanical properties and erosion resistance (Nayak 2020). In a similar study, areca sheath fibers modified using benzyl chloride were used as reinforcement (10–30%) for poly(vinyl alcohol) in an effort to increase the dry sliding abrasive behavior. A 4.2% improvement in grey relational grade was obtained when 30% reinforcement was used at a sliding velocity of 0.392 m/s. The wear rate of the composites was dependent on the sliding velocity and amount of reinforcement (Nayak 2017). Adding untreated and alkali (5%) and benzoyl-treated fibers obtained from areca sheaths improved the tensile, flexural, and impact strengths (Table 5.8). Fiber-reinforced PVA also had higher thermal resistance, shear modulus and loss modulus (Nayak 2019). At an optimum fiber content of 27%, areca sheath fiber-reinforced PVA composites had strength of 46 MPa and also provided better specific wear resistance (Fig. 5.6) and low coefficient of friction depending on the sliding rate and sliding distance. Hence, areca sheath fiber-reinforced composites were suggested to be suitable for applications where wear and friction resistance were necessary (Nayak 2019). In another study, the possibility of using areca sheath fibers for brake pad applications was studied. Fibers extracted from areca sheaths were combined (5 or 10%) with various other materials (Table 5.9), and composites suitable for brake pad applications were developed. Composites containing 10% fibers had high density and compressibility, and high fade and friction recovery. Comparatively, the composites with 5% fibers were porous and showed better wear resistance and frictional performance, suggesting suitability for brake pad applications (Table 5.10) (Krishnan 2019).

Similar to using PVA as resin, fibers extracted from areca sheaths after retting for 10 days were cut into 1–5 cm lengths and used as reinforcement for poly(vinyl chloride) (PVC, K57 grade). The fibers were further treated with 5% sodium hydroxide solution for 1 h at 80 °C and later immersed in benzoyl chloride for 15 min to reduce hydrophilicity. Treated fibers were combined with PVC in 0 to 40% ratio and made into composites. Treating fibers with alkali and benzyl chloride increased tensile strength from 66 to 115 MPa, modulus from 826 to 1294 MPa, and elongation from 2.6 to 3.2%. Substantial changes were also noticed in the tensile, flexural, and impact resistance of PVC reinforced with the treated fibers (Table 5.11). Incorporating the fibers into PVC also improved the biodegradability of the composites (Nayak 2020). In another study, areca sheath fibers were first treated with alkali (sodium hydroxide) and later functionalized with benzoyl peroxide in 5 to 8% concentrations. Later, the

Table 5.8 Properties of neat PVA and PVA reinforced with various levels of untreated and benzoyl chloride-treated fibers extracted from areca sheaths (Nayak 2019)

Composites	Fiber content, %	Tensile strength, MPa	Young's modulus, GPa	Elongation at break, %	Flexural strength, MPa	Flexural modulus, GPa	Impact strength (kJ/m^2)
	Neat PVA	12.26 ± 0.5	362.5 ± 26.8	1.73 ± 0.09	19.21 ± 0.86	2.3 ± 0.08	3.8 ± 0.21
Untreated	10	21.48 ± 0.44	875.73 ± 25.1	1.94 ± 0.10	22.10 ± 1.22	1.3 ± 0.10	3.3 ± 0.22
Treated	10	30.98 ± 0.62	1012.56 ± 31.0	2.59 ± 0.15	29.11 ± 1.66	2.6 ± 0.14	4.5 ± 0.31
	20	44.92 ± 1.05	1245.14 ± 42.8	2.84 ± 0.14	38.49 ± 1.89	3.0 ± 0.11	5.0 ± 0.27
	27	46.52 ± 1.02	1602.72 ± 48.6	2.91 ± 0.08	40.86 ± 1.27	3.4 ± 0.22	5.3 ± 0.18
	30	44.76 ± 1.38	1228.68 ± 51.1	2.41 ± 0.10	40.70 ± 1.79	2.6 ± 0.28	5.2 ± 0.20
	40	27.25 ± 1.42	1308.12 ± 56.0	2.25 ± 0.06	39.34 ± 1.20	2.4 ± 0.21	4.6 ± 0.30

Reproduced with permission from Taylor & Francis

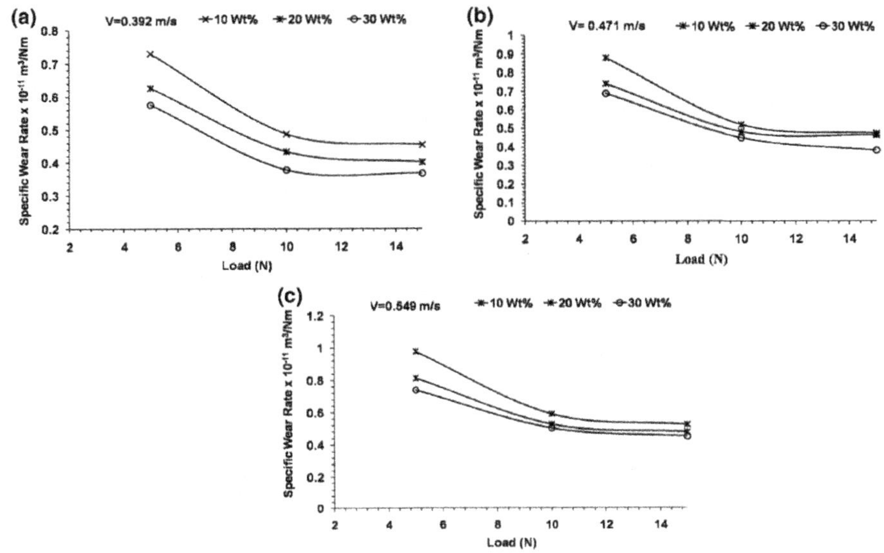

Fig. 5.6 Effect of load, sliding velocity, and amount of reinforcement on the specific wear rate of areca sheath fiber reinforced poly(vinyl alcohol) composites (Nayak 2019). Reproduced with permission from Taylor & Francis

Table 5.9 Composition of the areca sheath fiber reinforced composites intended for brake pad applications (Krishnan 2019)

Ingredients	Brake pad 1	Brake pad 2
Acrylic fiber	4	4
Rockwool fiber	6	6
Steel fiber	5	5
Areca sheath fibers	5	0
Hydrated lime	0	10
Phenolic resin	3	3
NBR	9	9
Crumb rubber	3	3
Calcium oxide	2	2
Silicon carbide	2	2
Artificial graphite	3	3
MoS_2	9	9
Exfoliated vermiculite	6	6
Barites	10	10
Friction dust	14	9
Mica	10	10
	9	9

Reproduced with permission from IOP Publishing

Table 5.10 Comparison of the frictional properties of areca sheath-based composites at two different ratios of areca sheaths (Krishnan 2019)

Properties	Unit	Standard	BPC1	BPC2
Density	g/cc	IS2742	2.19	2.13
Hardness	No unit	–	102	97
Acetone extraction	%	–	1.21	1.22
Compressibility	Mm	–	1.12	1.15
Porosity	%	JISD4418	4.7	5.1
Shear strength	N/mm^2	ISO6312	441	411

Reproduced with permission through IOP Science

treated fibers were combined with epoxy resin and formed into composites. Tensile strength of the composites was higher for alkali-treated samples and increased further when 4% benzoyl peroxide was used (Fig. 5.7). Further increase in concentration of benzoyl peroxide gradually decreases the strength. Similar results were also observed for the flexural strength (Kamath 2021). In another study, areca sheath fibers obtained after water and alkali retting were immersed in benzoyl chloride (4–8% in acetone) and used as reinforcement (55%) and compression molded at 40 bar pressure. Extent of increase in the strength of the composites was dependent on the concentration of benzoyl chloride. Highest tensile strength (4 MPa) and flexural strength (225 MPa) were obtained when treated with 4% benzoyl chloride (Kamath 2021).

In a similar study, the effect of tannin-glyoxal adhesive hardener on the cohesion and adhesion of areca sheath particle board was determined at different hardener

Table 5.11 Properties of neat PVC and PVC reinforced with various levels of untreated and alkali-treated fibers extracted from areca sheaths (Nayak 2020)

Composites	Fiber content, %	Tensile strength, MPa	Young's modulus, GPa	Elongation at break, %	Flexural strength, MPa	Impact strength, (kJ/m^2)
	Neat PVC	20.18 ± 0.38	1.57 ± 0.06	55.5 ± 0.08	10.6 ± 0.88	12.2 ± 0.22
Untreated	10	26.67 ± 0.87	1.95 ± 0.08	36.2 ± 0.10	12.2 ± 1.01	10.1 ± 0.28
	20	32.19 ± 1.01	2.17 ± 0.11	28.7 ± 0.09	17.6 ± 1.13	9.2 ± 0.19
	30	28.42 ± 1.11	2.36 ± 0.14	20.2 ± 0.12	15.7 ± 1.20	7.9 ± 0.17
	40	22.74 ± 1.18	2.59 ± 0.17	16.6 ± 0.06	10.3 ± 0.90	5.2 ± 0.07
Treated	10	34.74 ± 1.01	2.21 ± 0.04	44.3 ± 0.11	15.9 ± 1.28	10.3 ± 0.3
	18	42.38 ± 1.02	2.38 ± 0.09	40.62 ± 0.09	18.2 ± 1.09	9.1 ± 0.16
	20	40.79 ± 1.15	2.45 ± 0.11	32.3 ± 0.07	18.0 ± 1.17	8.5 ± 0.18
	30	37.03 ± 1.28	2.68 ± 0.13	29.1 ± 0.10	17.8 ± 1.08	8.2 ± 0.16
	40	25.66 ± 1.32	2.89 ± 0.15	22.5 ± 0.06	15.7 ± 1.13	7.8 ± 0.09

Reproduced with permission from Elsevier

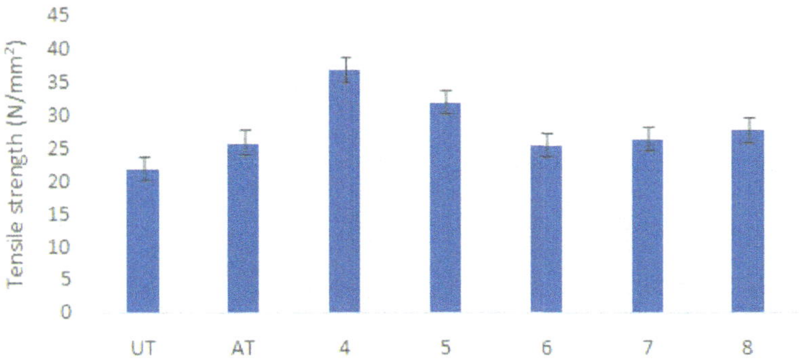

Fig. 5.7 Tensile strength of composites made from areca sheath fibers treated with benzoyl peroxide (Kamath 2021). Reproduced with permission through open access publication

levels in comparison to urea formaldehyde resins. Among the different adhesives studied, tannin-glyoxal provided maximum storage modulus and loss modulus with NH_4Cl as hardener. As seen from Fig. 5.8 the modulus of elasticity (MOE) and modulus of rupture (MOR) were higher for the adhesive formulation developed compared to commercially available urea formaldehyde. The tannin-based adhesives also provided better resistance to hydrolysis and lower water sorption (Anggini 2023). PVC composites reinforced with areca sheath fibers after chemical treatments also showed good resistance to solid particle erosion. Peak erosion was observed when the impact angle was 30° for pure PVC matrix compared to 60° for the fiber reinforced composites, suggesting that addition of areca sheath fibers induced ductility required for use in automotive and furniture applications (Nayak 2018).

Areca sheath was powdered and combined with urea formaldehyde resin and lignin or lignosulfonate as a flame retardant and later converted into composites. The resin, areca sheath, and lignin were combined in a drum mixer and later poured into a mold and heated at 150 °C for 10 min to form composites with density of 0.8 g/cm^3 (Fig. 5.9) (Madyaratri 2023). Lignosulfonate-containing composites had higher modulus of elasticity and water absorption compared to those containing lignin. Both the lignins provided particle boards with highest flame resistance rating of V0 and lowest mass loss of 16.7% (Madyaratri 2023).

Arecanut sheaths were immersed in water for 22 days and later dried and cut into lengths of 5-6mm. Cut sheaths were combined with ER 2074 epoxy resin and hand molded into composites with fiber content varying from 20 to 50% (Praveena 2022). Tensile strength of the composites varied from 20 to 50 MPa and modulus from 500 to 2000 MPa with 50% fibers providing the highest properties. SEM images (Fig. 5.10) showed strong interaction and bonding between the matrix and reinforcement with chemically treated fibers showing better interaction compared to the untreated fibers (Praveena 2022). Properties of the composites were considered to be suitable for marine, automotive, aerospace, and other applications. In order to further improve the properties, fibers obtained from areca leaf sheath were combined with polypropylene as matrix and maleic anhydride as the compatibilizer, and the mixture

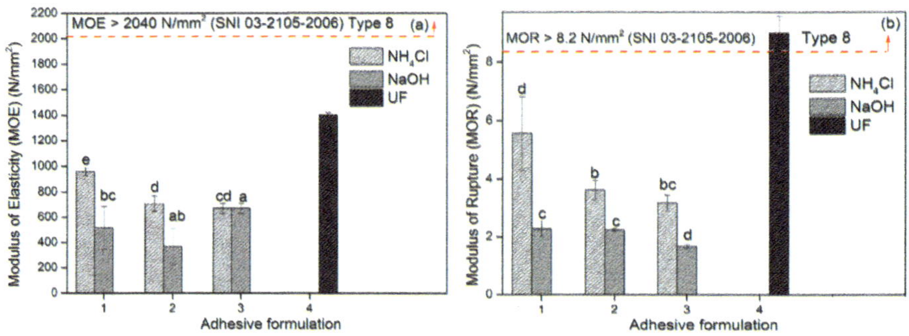

Fig. 5.8 MOE **a**, and MOR **b** of areca particleboard bonded with different TG adhesive formulations. The values with different letter **a–d** are significantly different (Anggini 2023). Reproduced with permission through open access publishing

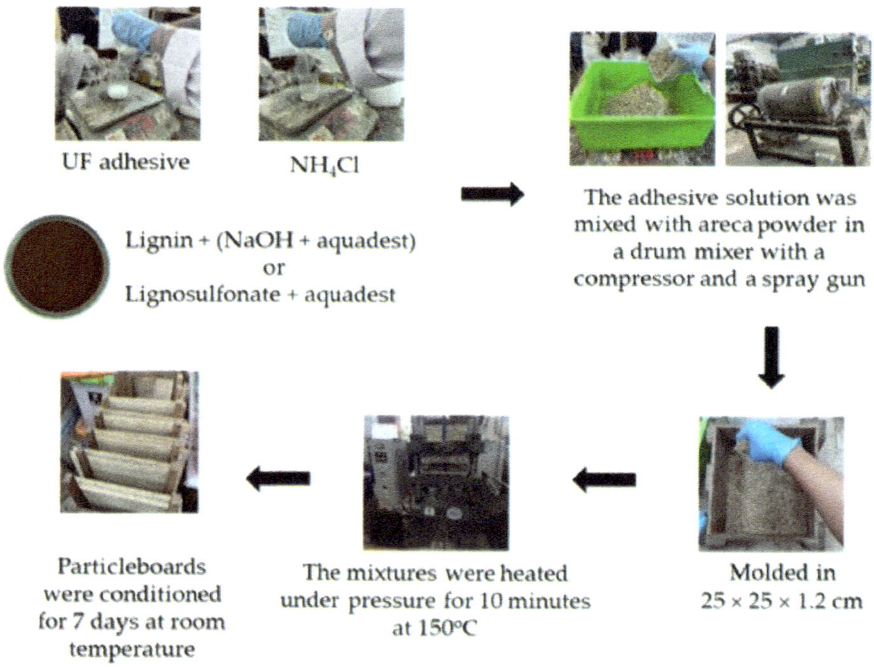

Fig. 5.9 Schematic of the process used to develop the areca leaf sheath composites with lignin and lignosulfonate as additives (Madyaratri 2023). Reproduced with permission through open access publication

was compounded in a twin screw extruder, and the pellets obtained were injection molded into composites. Addition of 5% fibers provided composites with highest strength and increasing ratio of the fibers further decreased the properties. However, presence of areca fibers increased water uptake and heat distortion temperature (Paul 2020).

Areca sheaths obtained from plants that were 3 to 6 years in age were treated with 5–15% of sodium hydroxide for 1 h and used as reinforcement for epoxy. The fibers were used in multiple layers and after treating with different concentrations of alkali. As seen from Figs. 5.11 and 5.12 and Table 5.12, the tensile strength and flexural strength of the composites were dependent on the number of layers, extent of treating in alkali, and the age of the areca sheaths (Figs. 5.11 and 5.12). Alkali treatment decreased hydrophilicity and also changed the morphological

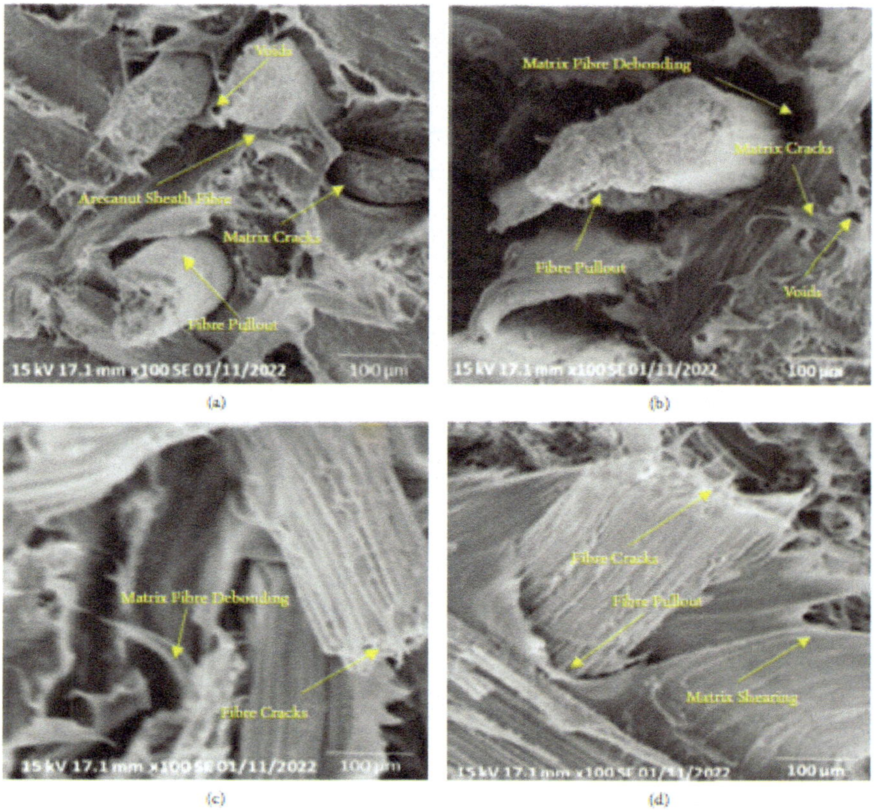

Fig. 5.10 SEM images depicting the interactions between the matrix and reinforcement at ratios of (20%-**a**, 30%-**b**, 40%-**c**, and 50%-**d**) (Praveena 2022). Reproduced with permission through open access publication

features promoting better interaction with the matrix due to microfibrillation leading to increased mechanical properties (Table 5.12) (Ashok 2020; Bangar 2020).

In a more recent study, fibers from areca sheaths were used as reinforcement for epoxy composites with bagasse ash and egg shell as fillers (5–15%). Properties of the composites were dependent based on the properties of the fillers, the amount

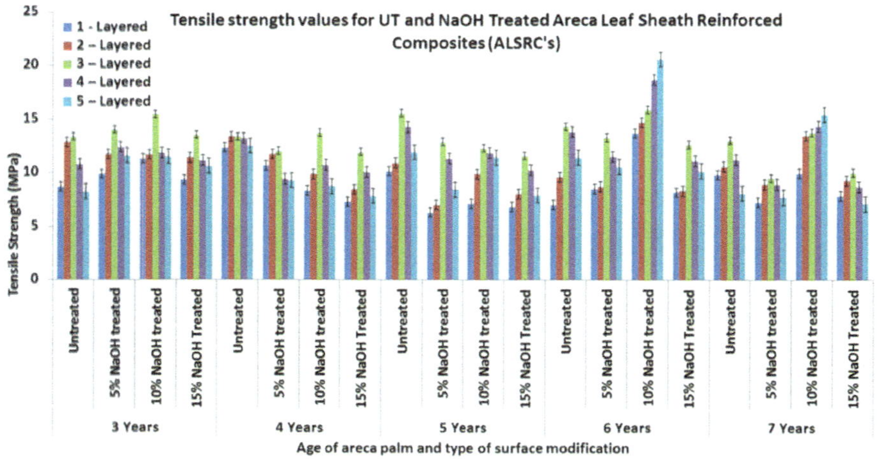

Fig. 5.11 Changes in the tensile strength of the composites reinforced with untreated and areca sheath fibers treated with various concentrations of alkali (Banagar 2020). Reproduced with permission from Taylor & Francis

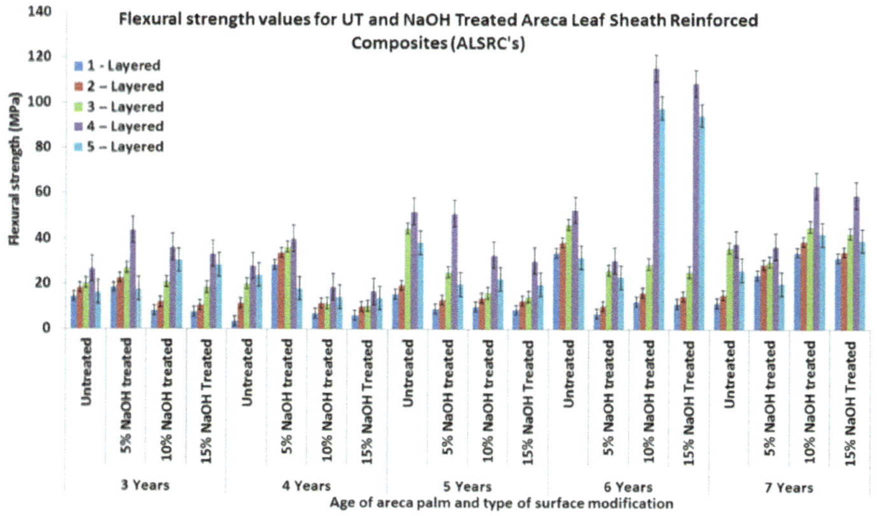

Fig. 5.12 Changes in the flexural strength of the composites reinforced with untreated and areca sheath fibers treated with various concentrations of alkali (Banagar 2020). Reproduced with permission from Taylor & Francis

Table 5.12 Effect of areca palm age, alkali treatment, and number of layers on the tensile strength (MPa) of the areca leaf sheath reinforced epoxy composites (Ashok 2020)

Age of palm, years	Treatment % NaOH,	Number of layers				
		1	2	3	4	5
3	0	8.65	12.81	13.35	10.75	8.20
	5	9.82	11.67	14.00	12.33	11.53
	10	11.29	11.64	15.43	11.79	11.45
	15	9.32	11.37	13.50	11.12	10.58
4	0	12.32	13.34	13.35	13.19	12.47
	5	10.62	17.70	11.99	9.35	9.26
	10	8.29	9.82	13.73	10.66	8.70
	15	7.29	8.42	11.89	10.01	7.82
5	0	10.11	10.88	15.52	14.23	11.88
	5	6.30	7.00	12.84	11.30	8.40
	10	7.12	9.83	12.25	11.80	11.38
	15	6.80	8.00	11.54	10.20	7.89
6	0	6.97	9.56	14.30	13.77	11.36
	5	8.46	8.68	13.24	11.44	10.52
	10	13.65	14.66	15.84	18.63	20.51
	15	8.16	8.28	12.58	11.02	10.11
7	0	9.76	10.52	12.99	11.18	8.07
	5	7.24	8.88	9.48	8.87	7.68
	10	9.91	13.43	13.73	14.29	15.39
	15	7.87	9.23	9.97	8.65	7.08

of fillers, and the extent of adhesion between the fibers and the matrix. Composite density increased with fiber loading, and void content increased for bagasse ash-filled composites but decreased for the egg shell-reinforced composites. Addition of fillers led to improved tensile, impact and dynamic mechanical strength with maximum tensile, flexural, and impact strength of 69.4 MPa, 121 MPa, and 34.8 kJ/m^2, respectively. Bagasse ash added to the composites provided the maximum storage and loss modulus, whereas egg shell reinforced composites provided the best damping properties. These composites were considered to be suitable for semi-structural applications or a substitute for common wooden materials (Joshi 2024).

In another study, fibers obtained from areca sheaths and palm leaf sheaths (Table 5.13) were combined in multiple layers and in different sequences and made into composites using epoxy as the resin. Considerable differences were observed in the tensile strength, flexural strength, and impact resistance properties depending on the configuration of the composites. Composites having a skin layer of areca sheath fibers and a core layer of palm sheath fibers had tensile strength of 46 MPa, flexural strength of 51 MPa, compression strength of 54 MPa, and shore hardness of 87, which was attributed to the better fiber–matrix interactions. Hybridization was suggested to be an ideal approach to obtain composites suitable for industrial applications (Ganesh 2020).

Fibers (Fig. 5.13) extracted by treating areca sheaths in water were combined with unsaturated polyester resin (ECMALON 4413, density of 138 g/cm^3) and made into composites. Composites were developed via the hand-lay-up technique with various fiber orientations and volume fractions. Thermal conductivity measurements on the composites revealed that the thermal conductivity was lower at the highest fiber volume fraction used and when the orientation of the fiber to the matrix was 0°. Comparatively, the highest thermal conductivity (0.215 W/m.K) was obtained at 90° fiber orientation. The thermal conductivity of areca sheath reinforced composites was suggested to be lower than that of bamboo and glass fiber reinforced composites, although the conductivity was not compared using similar composites or identical testing conditions (Gudapati 2018).

In another study, fibers were extracted from areca sheaths and converted into non-woven sheets (Fig. 5.14) for potential use as sound absorbers. Average diameter of the fiber obtained was 471 μm, density was 1.3 g/cm^3, roughness was between 0.141 and 0.180 mm, strength was 34 cN/tex, and elongation was 2.6%. Sheets (absorbers) of various thicknesses and density were made, and their ability to sorb sound was measured in terms of sound absorption coefficient using an impedance tube as per ASTM standard E-1050. The absorption coefficients of the areca sheets having different thicknesses are shown in Fig. 5.15. Absorption increased with increasing thickness, with an optimum of 54 mm (Raj 2020).

Hybrid Composites

In addition to binary blends, several studies have developed hybrid areca-based composites with other fibers and polymers. Multiple layers of areca sheath fibers were combined with glass fibers and/or jute fiber layers in several stacking sequences. Composites were developed by adding epoxy resin and hardener to the fibrous

Table 5.13 Properties of fibers obtained from areca sheaths and palm leaf sheath (Ganesh 2020)

Composition, %	Areca sheath fibers	Palm leaf sheath fibers
Cellulose	57.3	28.16
Hemicellulose	13.3	20.60
Lignin	22.0	31.8
Wax	0.2	5.0
Moisture	7.0	14
Tensile strength, MPa	254 ± 57	119 ± 38
Young's modulus, GPa	2.9 ± 0.6	1.8 ± 0.7
Elongation, %	11.9 ± 0.8	21.1 ± 0.5

Fig. 5.13 Fibers extracted from areca sheaths after retting in water for 5–7 days at room temperature (Gudapati 2018). Reproduced with permission through open access publication

Fig. 5.14 Images of the areca palm tree and its parts are used for developing the acoustic sound absorbers (Raj 2020). Reproduced with permission from Elsevier

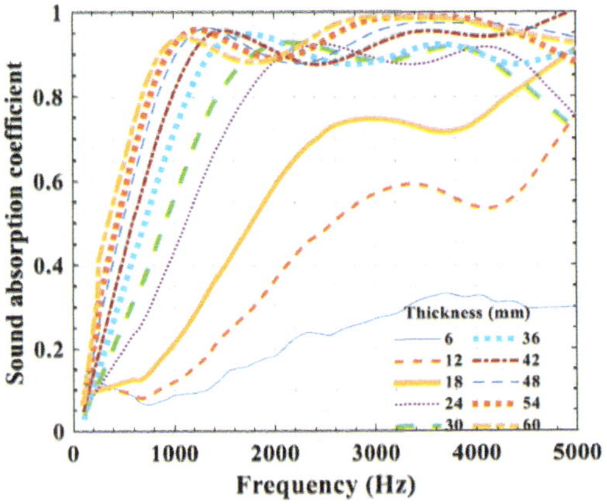

Fig. 5.15 Effect of thickness of the areca sheath fiber mat on the sound absorption coefficient (Raj 2020). Reproduced with permission from Elsevier

reinforcements (Jothibasu 2020). High-strength glass fibers in longitudinal, skin/ intermediate layer provided higher strength to the composites. Similarly, alkali-treated sheath fibers provided composites with better properties compared to untreated composites. The composites were considered suitable for manufacturing furniture and other applications.

5.7 Reinforcement for Concrete

Similar to composites, fibers obtained from areca sheaths have been used as reinforcement for concrete. In a study by Hasan et al., areca sheath fibers having length of 50 mm and diameter of 0.75 mm were obtained by treating them in water. These fibers were combined with concrete with fiber ratios between 0 and 4%. Addition of the fibers drastically reduced the slump value and density of the concrete. However, compressive strength of the concrete showed an increase when 1% fibers were used, but decreased at higher fiber content. Higher strength of the 1% fiber containing concrete was retained even after 28 days of curing. Splitting strength of concrete was also enhanced by up to 18%, and concrete showed better resistance to cracking. Addition of areca fiber was considered to be beneficial in terms of better cost index, lowering carbon intensity, and reduced Ca:Si ratio (Fig. 5.16) (Hasan 2023).

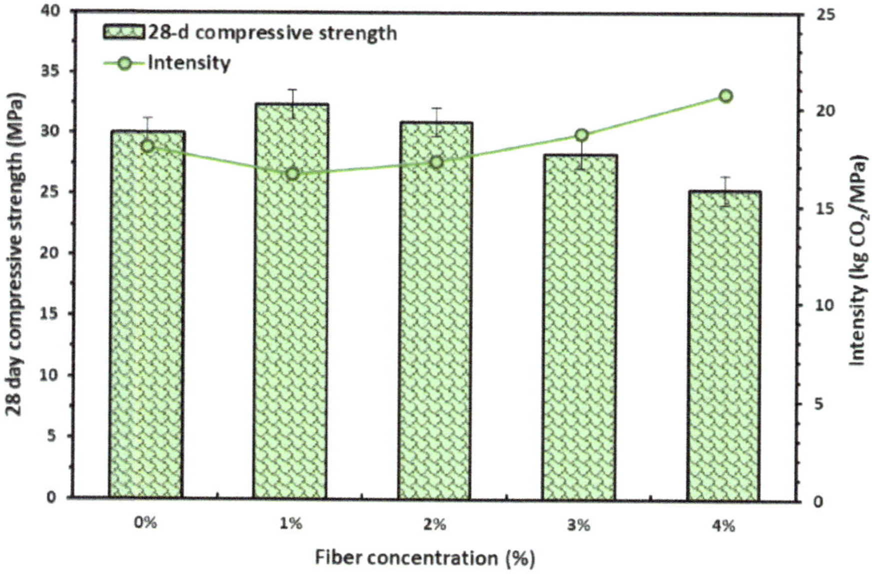

Fig. 5.16 Eco-strength efficiency of concrete reinforced with areca sheath fibers with respect to compressive strength (Hasan 2023). Reproduced with permission through open access publishing

5.8 Areca Leaf Sheath Compost

One of the simplest and natural approaches to dispose areca sheaths is by on-site composting. Although composting could provide some nutrients to the soil, the process of natural decomposition is considerably longer and not suitable for commercial areca cultivation. Also, prolonged degradation periods will increase the exposure of the sheaths to microbial attacks, which could become an issue for cultivation and long-term growth of the areca plants. Hence, it is necessary to remove areca sheaths from the field and find alternative uses. Developing compost from areca sheaths and other areca wastes at off-site locations is considered to be a prudent approach to treat the byproducts. Areca sheaths were made into pieces and combined with either vermicompost or soil and effective microorganisms. The compost developed was added in various proportions for the growth of *Vignaunguiculata* L., and the changes in the plant growth were observed. Protein, chlorophyll, and carbohydrate contents in the plant increased substantially due to the addition of the areca sheath composts (Maheshwari 2015; Bhat 2024).

5.9 Converting Areca Sheaths into Packaging Materials

Areca sheaths are one of the most unique materials available in nature. In addition to being byproducts that are renewable and biodegradable, the sheaths have good strength, durability, and resistance to moisture. The sheaths are considerably flexible and can be molded into different shapes and sizes (Fig. 5.17). Hence, areca sheaths are extensively processed into plates, cups, bowls, and many other shapes for packaging applications. Typically, areca sheaths are directly made into various products by simple compression using either completely manual or semi-automatic machines that have been developed to compress the sheaths into various products. Figure 5.18 shows a semi-automatic machine that is used to compress the sheaths into plates and cups. Different shapes and sizes of areca sheath products are obtained by changing the molds. A detailed investigation on the compressibility and deforming ability of the areca sheaths was studied by Mohanty et al. Areca sheaths were cut into required dimensions and subject to uniaxial tensile testing, compression, and plane-strain rolling. Areca sheaths were found to be capable of withstanding strains between 0.5 and 2 during stretching, compression, or rolling. Deformation limits were better when the sheets were hydrated (Fig. 5.19). Areca sheath-based products are considered to have the potential to replace several synthetic polymer-based products and have shown to have positive life cycle analysis, and could help to reduce the waste management and emission issues related to using plastic-based products (Gautam 2017). Conditions, particularly temperature, used during the conversion of areca sheaths into packaging were reported to have a significant influence on the appearance and usability. An optimum temperature of 170 °C, compression time of

2 min, and moisture level in the sheath between 20 and 60% (Fig. 5.20) was found to be the most optimum for preparing food containers (Prasan 2025).

Both manual and semi-automatic machines require considerably human effort and energy to compress the sheaths into various shapes and sizes. To reduce the burden and improve efficiency, a novel hand-operated machine was developed to

Fig. 5.17 Some of the cutlery products developed from areca sheaths

Fig. 5.18 Automatic areca leaf pressing machine

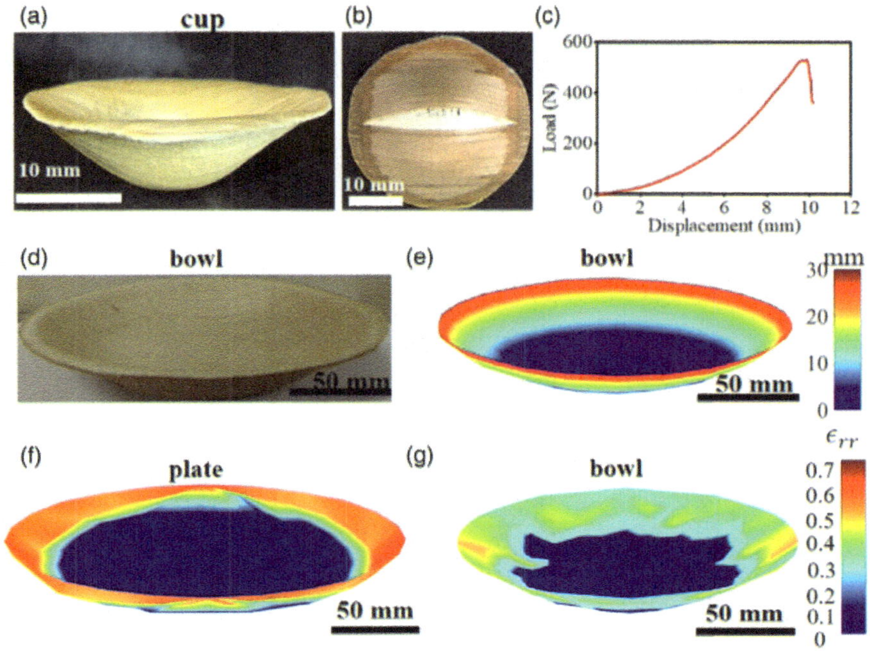

Fig. 5.19 Digital images of the areca cups compressed during the LDH test **a**, failure region on the back of the cups **b**, load displacement curve **c**, digital image of the bowl **d**, reconstructing the bowl shape **e**, radial strain distribution in plate **f** and bowl **g** (Mohanty 2021). Reproduced with permission from Springer Nature

Fig. 5.20 Image of areca sheaths having moisture content of 20%, 20–60%, and higher than 60% processed into plates (Prasan 2025). Reproduced with permission through open access publishing

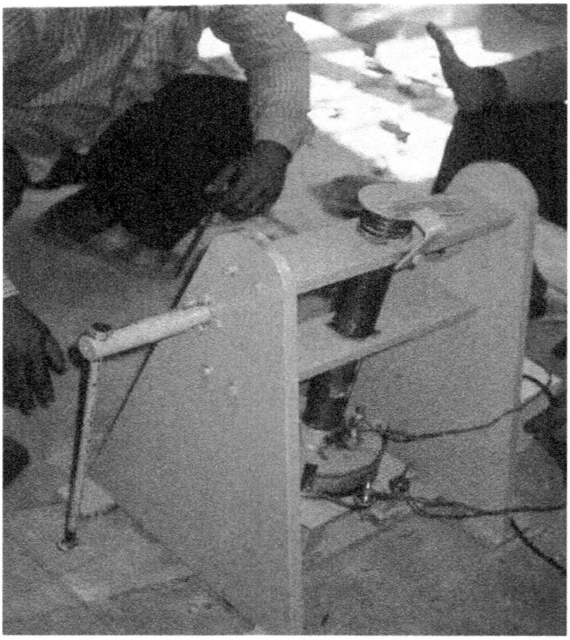

Fig. 5.21 A hand-operated machine for the conversion of areca sheaths into various products (Kalita 2008)

convert areca sheaths into various products (Kalita 2008). The improved machine could be operated by hand in the seating position making it possible for physically disabled to operate the machine. Time required to produce a bowl was reduced by 50%, and that for production of plate was reduced by 30%. A punching temperature of 110–115 °C, die temperature of 35–45 °C, and moisture content in the sheath between 14 and 17%, and compression time of 40-60s was considered optimum. The new machine (Fig. 5.21) was reported to cost at least 50% less than existing foot-operated machines.

5.10 Preserving Areca Sheath-Based Packaging Materials

Despite considerable efforts to utilize areca sheaths as a source for packaging materials, there have been several limitations. Being biobased materials, areca sheaths have inherently high (14–20%) moisture absorption and, under high humidity conditions, are susceptible to bacterial contamination (Matan 2011). Several approaches have been used to improve the resistance of the areca sheaths to bacterial and fungal contaminations. In one such study, areca sheaths were treated with various essential oils, and the ability of the oils to inhibit microbial growth was studied. It was found that cinnamon oil with a minimum inhibitory concentration of 50 μg/ml had better activity against microbes and was able to protect the sheaths against molds for up to 12 weeks when stored at 25 °C and 100% relative humidity. It was suggested that cinnamaldehyde in cinnamon oil was responsible for the high antimicrobial activity and suitable for protecting the sheaths from microbial degradation when used for packaging intermediate moisture foods (Matan 2011). Similar to cinnamon oil, clove oil, and eugenol were able to substantially inhibit growth of microorganisms on areca sheaths. It was also found that treating the sheaths with clove oil or eugenol under cold atmospheric plasma (40 W, 10 min) was able to enhance the activity against *A. niger, Penicillium sp.*, and *Rhizopus sp.* All three microorganisms were inhibited with clove oil concentration of 30 μl/ml and eugenol concentration of 10 μl/ml (Fig. 5.22). Treating with plasma was able to increase the effectiveness (reduce MIC) by up to 10 folds at 25 °C and 100% humidity for up to 12 weeks. It was suggested that although eugenol in clove oil was responsible for the antifungal activity, plasma treatment was necessary to obtain adequate inhibition for food applications (Matan 2014). Plasma treatment created UV species, free radicals, and charged particles, which decreased hydrophilicity and improved the bacterial resistance of the sheaths (Matan 2014). In another study, it was reported that applying the first, second, and third extracts from crude coffee and crude spent coffee grounds was able to inhibit the growth of *Aspergillus sp., Aspergillys flavus, Aspergillus niger, Pencillium citrinum, Penicillum chrysogenum, and Eurotiumams stelodami.* Areca sheaths were treated with the extracts and used to wrap durian paste. The extracts had minimum inhibitory concentrations between 100 and 230 μg/ml for the first extracts and 300 to 460 μg/ml for the second extracts. The shelf life of the packaged food was increased from 3 to 21 days when stored at 25 °C and 100% humidity. Caffeine (80–85%), phenol, and ferulic acid were the constituents that were responsible for providing the antimicrobial activity (Nonthakaew 2015).

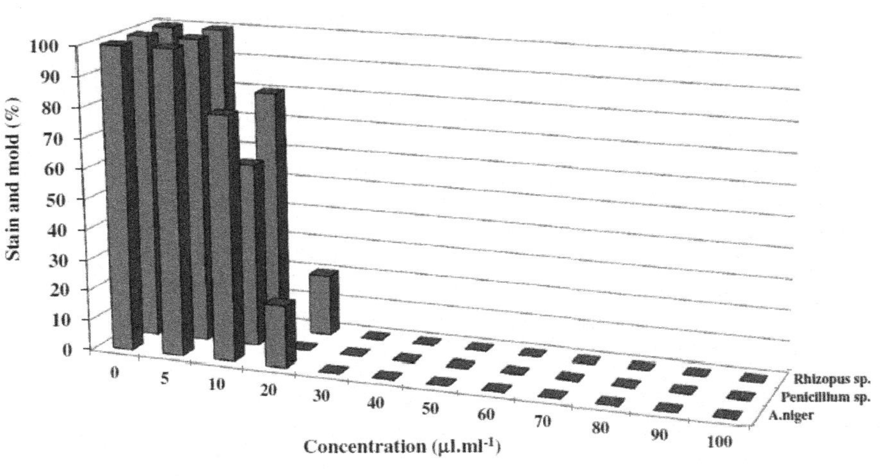

Fig. 5.22 Decrease in the stain and mold size of fungi on areca sheaths treated with clove oil and plasma after incubation at 25 °C and 100% humidity for 12 weeks (Matan 2014). Reproduced with permission from Elsevier

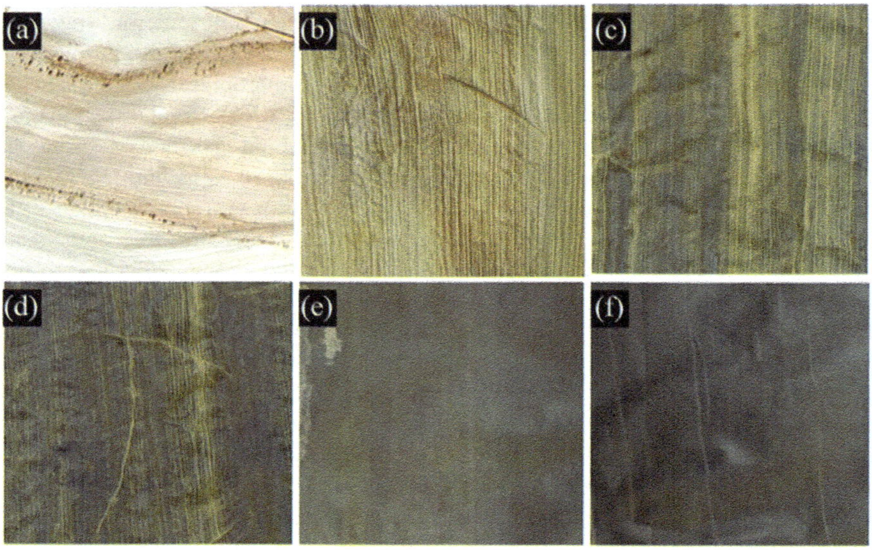

Fig. 5.23 Digital images of the areca leaf sheath after different treatments. Raw untreated **a**, unbleached after initial processing **b**, sample after 40 min bleaching **c**, Sample after 50 min bleaching **d**, transparent sample after 60 min bleaching (Dissanayake 2021). Reproduced with permission from Springer Nature

In order to improve the transparency, mechanical properties, and enhance the suitability for food packaging, areca leaf sheaths were immersed in water for 48 h, and layers were manually peeled. The peeled layers were opaque and hence treated with hydrogen peroxide and heated at 90–95 °C for 60 to 90 min. Considerable visual and mechanical changes occurred with the layers turning transparent (Fig. 5.23), similar to synthetic polymer films used for food packaging. A layer of carboxy methyl cellulose (CMC) was applied to improve mechanical properties. However, even at higher density and thickness, the coated and uncoated leaf sheath layers had poor properties compared to polythene used for packaging (Table 5.14) (Dissanayake 2021).

5.11 Production of Enzymes

Efforts have been made to understand the possibility of using areca sheaths as substrate for enzyme production. Areca sheaths were powdered and incubated with *Pseudomonas fluorescens,* and solid-state fermentation was done to produce cellulase. Amount of enzyme produced was dependent on the conditions, such as incubation time, pH, fermentation media, etc. Under the optimized condition, an enzyme activity of 0.942 U/ml was possible (Shashidhar 2013).

Table 5.14 Comparison of the properties of the uncoated and CMC-coated areca leaf sheaths (ALS) with polythene (Dissanayake 2021)

Property	Polythene	ALS uncoated	CMC-coated ALS
GSM	46.4	112.3	132.5
Breaking force, N			
– Lengthwise	57.31	7.78	12.78
– Widthwise	57.31	35.03	40.33
Mean elongation at break, mm			
– Length wise	150	12.16	11.26
– Width wise	150	9.04	10.14
Tearing strength, g			
– Length wise	452	201	320
– Widthwise	453	202	319
Bursting strength, kg/cm^2	0.1	0.1	0.3
Thickness, mm	0.5	0.85	1.0

5.12 Bioethanol from Areca Sheaths

The high cellulose content (up to 73%) was considered to be suitable for converting into ethanol. Sheaths were pretreated with 5% nitric acid or alkali (10% NaOH) and later subject to enzyme hydrolysis and subsequent fermentation to form ethanol. Pretreatment resulted in the breakdown of lignin, and the cellulase enzymes used were able to hydrolyze the cellulose into glucose, where sugar levels between 16 and 25 mg/ml were obtained. *S. cerevisiae* was able to successfully ferment the sugars into ethanol within 7 days. Refractive index and bioethanol content obtained after various pretreatment levels from the areca sheaths are given in Table 5.15 (Wahyuti 2025).

Table 5.15 Refractive index and bioethanol content obtained after fermentation of areca sheaths pretreated to various extents (Wahyuti 2025)

Pretreatment	Fermentation days	Refractive index	Bioethanol content, %
HNO_3	3	1.3362	12.00
HNO_3	5	1.3363	12.25
HNO_3	7	1.3381	16.75
NaOH	3	1.3355	10.25
NaOH	5	1.3373	14.75
NaOH	7	1.3363	12.25

5.13 Absorbent for Heavy Metal Removal

Areca leaf sheaths were ground to pass through an 80 size mesh and later chemically treated with concentrated sulfuric acid for 5 h at 60 °C in order to increase the surface area and microporosity. The modified sheath powder was used as a sorbent for Cr(VI) at various sorption conditions. Concentration of the chromium before and after sorption was measured using ICP-MS (Pant 2022). pH2 was found to provide the most optimum absorption and chemically modified sheaths had 69% removal efficiency compared to only 17% for the unmodified sheath powder. Similarly, initial concentration and sorption time also affected the level of absorption. Highest sorption of 109 mg/g was obtained which was higher than many other biosorbents previously used. Adsorption was favored through pseudo-second order kinetics and followed Langmuir model.

Fig. 5.24 Photographs of the areca sheath and other materials used for the soil stabilization studies (Kolathayar 2019). Reproduced with permission from Taylor and Francis

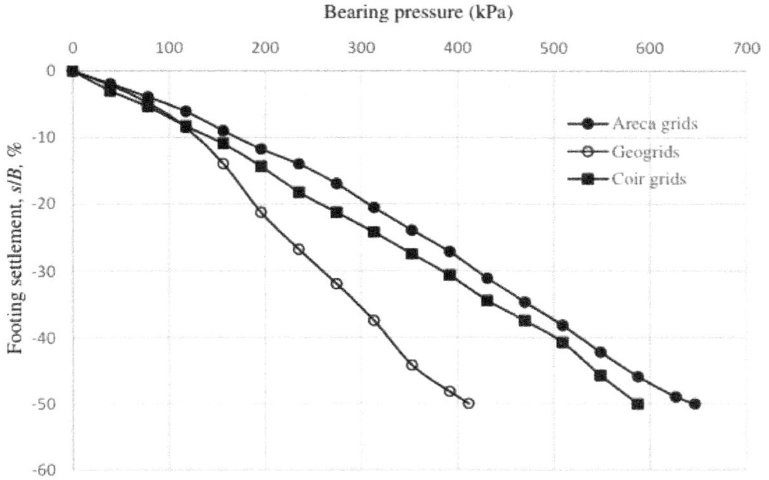

Fig. 5.25 Comparison of the bearing pressure-settlement behavior of soil reinforced with areca geogrids compared to coir and synthetic geo grids (Kolathayar 2019). Reproduced with permission from Taylor and Francis

5.14 Reinforcement for Soil

Areca sheaths typically take 1 to 2 years to degrade in normal conditions and were hence considered to stabilize soil. Thin strips of about 5mm in width were separated from the sheaths and hand-woven into a grid similar to that of coir and synthetic geo grids Fig. 5.24. Areca sheath grids had highest tensile strength of 44 kN/m substantially higher than that of commonly used geo grid (10 kN/m). Sand beds reinforced with the areca sheath geo grids showed up to 3.7 times increase in bearing pressure and 1.6 times higher than that compared to the synthetic geo grid (Fig. 5.25) (Kolathayar 2019).

5.15 Substrate for Mushroom Cultivation

Areca sheaths, particularly those available as waste after manufacturing of plates, were treated as a substrate for cultivation of edible mushrooms *Pleurotus ostreatus* spp in combination with paddy straw or orange peel in different ratios. A blend of 1:1 ratio of arecanut, paddy straw, and orange peel provided the best results with highest rate of primordial formation, highest average number of fruiting bodies, the greatest average weight of individual fruiting bodies, and also yield. In terms of nutritional composition, mushrooms obtained using this combination provided

3.2 g of protein per 100 g of mushrooms and accounted for about 32 cal. It was suggested that areca sheaths could be an economical alternative to grow mushrooms and provide the desired properties (Monisha 2024).

5.16 Conclusions

Areca sheaths are unique byproducts with distinct structure and composition. Good mechanical strength, high resistance to aqueous environments make them preferable as natural packaging materials. Areca sheaths are converted into plates, cups, and several other forms for food and non-food packaging on a commercial scale. Attempts have also been made to convert the sheaths into pulp for paper production, as a source for nanocellulose and lignin. Some studies have also been done on using areca sheaths as reinforcement for composites and concrete. Unlike the husk or the nuts, there have been limited studies on utilizing areca sheaths. Studies to convert the sheaths into carbon, as substrate for production of enzymes, mushrooms etc., could be considered.

Chapter 6
Areca Leaves and Trunk

6.1 Energy Applications of Areca Leaves

Areca plants shed the leaves regularly and form the major residue if weight of the byproducts per tree is considered and are hence available in large quantities. To ensure appropriate use, the possibility of converting the leaves into briquettes for energy applications was studied (Deepak 2013). Leaves were dried and cut into sizes of 1700 μm or longer. Initial studies showed that compaction of the leaves alone did not provide stable briquettes. Hence, the leaves were combined with paper, saw dust, or coconut coir powder as binders (in 2:1 ratio) and compacted in a hydraulic press to form bricks (Fig. 6.1). The composition and calorific properties of the briquettes are shown in Table 6.1. In addition to large availability at low cost, the areca leaf briquettes were found to be capable of removing carbon and sulfur more efficiently than common biomass sources (Deepak 2013). Areca leaves (powdered to 850 μm mesh size) could also be combined with saw dust and when made into briquettes, had a calorific value of 14.57 MJ/kg and low sulfur (0.65%) and nitrogen (0.61%) content, making them suitable as biomass fuel for both domestic and industrial applications (Deepak 2019). In another study, areca leaves shredded to about 600 μm mesh size were combined with saw dust and wheat flour and used as a binder to develop briquettes of various densities. The briquettes obtained under the optimum conditions had a gross calorific value of 3.3 kCal/Kg considered sufficient to generate heat and suitable for use as a biofuel (Deepak 2015).

Similar to the other studies on using the byproducts of areca plants after converting them into carbon, a study was done to carbonize areca leaves and understand the ability of the carbon to sorb dyes in water. Areca leaves were subject to carbonization at 300 °C and also activated using 10% $ZnCl_2$ solution or irradiated in a microwave for 60 min at a microwave power of 90 W and 270 J/sec. Ability of the different carbons to sorb crystal violet or congo red under varying sorption conditions was studied (Basrur 2019). Thermodynamic and isotherm parameters for the three carbons were comparable with sorption of crystal violet between 27 and 28 mg/g and congo red

N. Reddy and P. Aramwit, *Areca catechu and its byproducts*,
https://doi.org/10.1007/978-3-032-22134-6_6

Fig. 6.1 Briquettes made by combining areca leaves with coconut coir (left), saw dust (center), and paper (right) (Deepak 2013). Reproduced with permission through open access publishing

Table 6.1 Composition and properties of biomass briquettes prepared from areca nut leaves and paper in comparison to saw dust and coir powder as binders (Deepak, 2013). Reproduced with permission through open access publishing

Parameter	Areca leaves and Paper	Sawdust	Coconut coir
Gross calorific value, kcal/kg	2969	2680	3392
Moisture content, %	6.21	6.18	5.73
Ash content, %	8.24	8.30	9.23
Volatile matter, %	81.58	81.58	81.24
Fixed carbon, %	3.97	3.94	3.80
Hydrogen, %	6.97	6.97	6.94
Nitrogen, %	0.47	0.47	0.48
Sulphur, %	0.60	0.61	0.60
Oxygen, %	19.65	19.42	19.30

being 20–26 mg/g. Absorption of the dye onto the carbon was suggested to be due to the C = C bonds and also because of π-π interactions and hydrogen bonding. Greater affinity was found for the cationic crystal violet than the anionic Congo red due to the basic groups on the carbon (Barsur 2019).

Areca leaves were made into carbon by treating at 600 °C for 1 h and later activated with KOH with weight ratio of 1:2 and further carbonizing at 800 °C for 1 h for potential use in developing symmetrical supercapacitors (Le 2020) (Fig. 6.2). The carbon obtained was washed with 1 M HCl and water to further increase surface properties. Activation and treating with HCl substantially increased the porosity and pore size with highest activation energy of 876 m^2/g. Carbon obtained after treating with KOH and HCl was considerably amorphous and had low degree of graphitization. Further, the surface of the carbon had high level of oxygen-containing functional groups, which provided higher capacitance and wettability and hence better CV characteristics for the electrodes. A symmetrical supercapacitor developed using carbon showed high specific capacitance of 132 F/g at 0.5 A/g and energy density of 10.3 Wh/kg and power density of 375 W/kg. The supercapacitor had high durability with 92% of performance retention even after 5000 cycles (Le 2020). In a recent study, the feasibility of using areca leaf ash-K_2CO_3 as a heterogenous catalyst for production of biodiesel was evaluated (Boro 2024). For this application, areca leaves were converted into ash by open-air burning, and the ash obtained was mixed with 20% K_2CO_3, and the mixture was calcined at 550 °C for 1 h and used as a catalyst for the transesterification process for production of bioethanol (Fig. 6.3). Elemental composition of the catalyst obtained is given in Table 6.2. With a mesoporous structure and high surface area of 17.77 $m^2\ g^{-1}$, the catalyst was able to provide a biodiesel yield of 96.55 and conversion efficiency of 97.99%. Further, the catalyst could be reused for up to three cycles without affecting biodiesel yield (82.8%).

Activated carbon from areca leaves could also function as electrocatalyst for generation of H_2 from water through water electrolysis (Palanichamy 2013). In this

Fig. 6.2 The leaves of areca palm trees are used for preparing the carbon and the performance of a supercapacitor made using the KOH-activated and HCl-treated carbon made into a symmetrical supercapacitor (Le 2020). Reproduced with permission from Springer Nature

Fig. 6.3 Schematic of the process used to obtain the areca leaf catalyst for production of biodiesel (Boro 2024). Reproduced with permission from Elsevier through open access publishing

Table 6.2 Elemental composition of uncalcined areca nut leaf ash/K_2CO_3, calcined areca nut leaf ash/K_2CO_3, and third recycled areca nut leaf ash/K_2CO_3 catalysts and areca nut leaf ash catalyst (Boro 2024). Reproduced with permission from Elsevier through open access publishing

Elements	Composition of the catalysts							
	Uncalcined areca nut leaf ash/ K_2CO_3		Calcined areca nut leaf ash/K_2CO_3		3rd recycled areca nut leaf ash/ K_2CO_3		Areca nut leaf ash	
	Weight %	Atomic %	Weight %	Atomic %	Weight %	Atomic %	Weight %	Atomic %
C	25.85	33.15	23.68	34.41	47.17	59.36	38.40	48.98
O	44.14	48.15	42.41	46.27	30.89	29.18	43.16	41.32
Na	0.48	0.36	0.92	0.70	1.36	0.89	0.52	0.35
Mg	1.19	0.85	2.40	1.73	0.71	0.44	1.02	0.65
Al	0.21	0.13	–	–	2.89	1.62	–	–
Si	19.69	12.22	17.69	11	13.11	7.06	13.49	7.36
P	0.97	0.55	1.19	0.67	0.25	0.12	–	–
S	0.23	0.12	0.20	0.11	–	–	–	–
Cl	–	–	0.31	0.15	–	–	0.48	0.21
K	8.03	3.58	8.36	3.73	2.19	0.85	2.12	0.83
Ca	2.21	0.96	2.81	1.23	0.89	0.34	0.81	0.31
Fe	–	–	–	–	0.53	0.14	–	–

process, areca leaves were powdered and combined with $ZnCl_2$ in 1:1 ratio and later pyrolyzed at 250 °C for 2 h under atmospheric conditions. Later, the activated carbon (30% yield) was reduced using $NaBH_4$ as the reducing agent to obtain 10% Ni and 1% Pt supported biocatalysts. Electrodes were prepared using the areca carbon-based biocatalysts, and electrochemical studies were performed. No change in current density was observed for the bare or the electrode without the modified carbon. Distinct evolution of hydrogen was observed in the 1% Pt supported carbon due to the oxidation of organic moieties. The activated carbon without the supports was a good conductor and could also be suitable for sorption of heavy metals. It was suggested that the areca leaf carbon could become an ideal source for generation of H_2 gas for clean energy and other applications (Palanichamy 2013).

6.2 Applications of Areca Leaf Extracts

6.2.1 Medicinal Applications

Areca leaves are reported to contain highly valuable components suitable for treating various illnesses in humans and animals, as an additive to improve performance of materials, and as raw material for synthesis of various chemicals and compounds. For example, petroleum ether, chloroform, or methanol were used to extract the components in areca leaves, and the extracted components were studied for their anti-glycemic or antidiabetic properties. The extracts were administered to STZ-induced diabetic rats. All the extracts showed anti-glycemic activity with prominent reduction in elevated glucose levels. Extracts obtained using methanol were more active compared to ether or chloroform (Table 6.3). Not only reduction in glucose but also changes in body, liver, and kidney weight were also noticed (Mondal 2012). In another study, it was found that methanol extracts from areca leaves had anti-hyperglycemic activity, and the extent of activity was dependent on the dosage level. When injected into mice at 400 mg/kg, an 65.7% glucose inhibition compared to 71.4% for standard drug at 10 mg/kg dosage was observed (Akhter 2014).

Table 6.3 Extent of reduction in fasting blood glucose levels after injecting with extracts obtained from areca leaves using three different chemicals (Mondal 2012). Reproduced with permission through open access publishing

Group	Dose	Day 0	Day 5	Day 10	Day 15
I normal saline	5 ml/kg	76.9 ± 2.1	75.3 ± 4.2	74.9 ± 3.6	74.7 ± 2.9
II (STZ)	50 mg/kg	278 ± 9.2	281 ± 10.5	286 ± 11.4	293 ± 9.8
III (STZ + PET ether extract)	200 mg/kg	261 ± 11.9	111 ± 7.8	91 ± 1.6	88 ± 1.8
IV (STZ + methanol extract)	200 mg/kg	266 ± 16.5	91 ± 4.4	84 ± 3.2	76 ± 2.9
V (STZ + methanol extract)	200 mg/kg	272 ± 10.3	93 ± 3.7	83 ± 1.8	75 ± 2.6
VI (STZ + methanol extract)	0.5 mg/kg	279 ± 13.3	95 ± 2.8	82 ± 1.2	73 ± 2.4

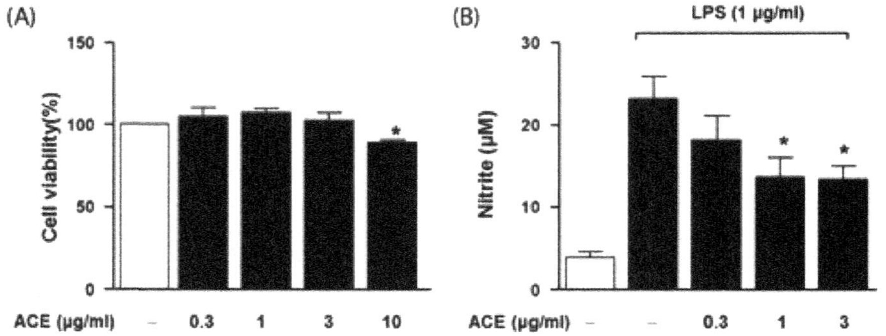

Fig. 6.4 Effect of Areca catechu leaf ethanol extract (ACE) on cell viability and production of nitric oxide (NO) in lipopolysaccharide (LPS)-stimulated RAW 264.7 cells. **a** RAW 264.7 cells were treated with 0.3–10 μg/ml ACE for 24 h, and cell viability was determined with the XTT assay **b** RAW 264.7 cells were treated with 0.3–3 μg/ml ACE and 1 μg/ml LPS for 24 h. NO production was determined with the Griess reagent assay. Data were expressed as mean ± SD. *indicates a significant difference compared with LPS treatment only ($P < 0.05$) (Lee 2014). Reproduced with permission through open access publishing

In addition to reducing glycemic levels, the possibility of using areca extracts as anti-inflammatory compounds was investigated. Areca leaves were treated with ethanol (70%) for 3 days at room temperature. Extracts obtained were washed and lyophilized, and the powder obtained was collected for further analysis and the anti-inflammatory properties of the extract were studied using LPS-induced RAW 264.7 macrophages and also using rat inflammation models. It was found that the areca leaf extracts showed significant ACE inhibited inflammation by regulation of the NF-kB expression by suppressing IkB phosphorylation. Similar results were also obtained when in vivo studies were done using mice through cell inflammatory response (Fig. 6.4) (Lee 2014).

6.2.2 Anticorrosive Applications

In a different approach, it has been shown that the extracts obtained from areca leaves can inhibit corrosion of aluminum. Four different concentrations (1, 2, 4, and 6 g/L) of areca nut leaf inhibitor were obtained by treating the leaves with acetone in a Soxhlet extractor for 7 h. Cleaned aluminum was placed in 0.5 M HCl solution with and without the extract, and the weight loss and changes in electrochemical properties were determined (Raghavendra 2018). The areca leaf extracts provided considerable protection to the surface and electrochemical properties of aluminum against degradation by the acid. As seen from Fig. 6.5, the surface of the metal without the extract was highly corroded, whereas the grain structure and surface morphology were preserved when the extracts were present. A surface roughness of 323 nm was obtained for the untreated metal compared to 186 nm for the metal where the corrosion

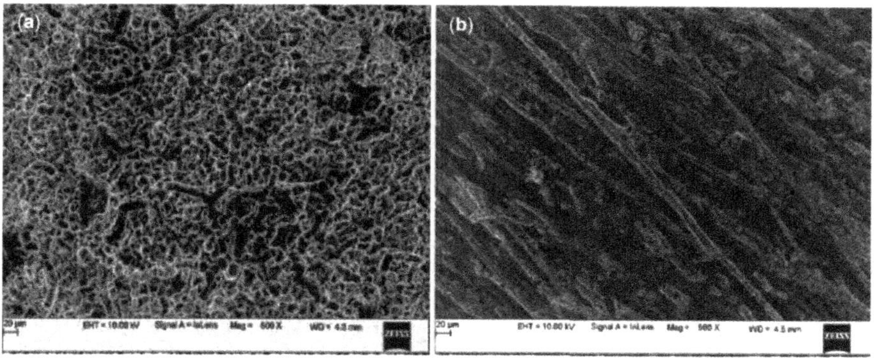

Fig. 6.5 Difference in the morphology of the aluminum after immersion in 0.5 M HCl without (left) and with the areca leaf extract (Right) (Raghavendra 2018). Reproduced with permission through open access publication

was inhibited by the areca extracts. Up to 97% protection efficiency was possible when 6 g/l of the extract was used, and the protection was mostly due to the charge transfer phenomenon (Raghavendra 2018). In another study, extracts obtained from mature arecanut husks had a density of 1.009 g/ml, viscosity of 0.877 centipoises, and surface tension of 86.8 dynes/cm and showed the presence of hydroxyl and carbonyl groups. This extract was added to copper and mild steel in both acid (0.5 M HCl) and alkaline (0.5 M NaOH) media, and the changes in the surface morphology and corrosion inhibition of the metals were studied at different temperatures. The extract was able to form a monolayer on the metal surfaces and acted as an anodic inhibitor for copper and a mixed inhibitor in other conditions. Inhibition was mainly through the formation of films on the surface, and the maximum protection was obtained when 18 g/L of the extract was used and immersion time was 1 h at 30 °C (Raghavendra 2018).

6.2.3 Nanoparticle Synthesis

Aqueous extracts from areca catechu leaves were used for synthesis of silver nanoparticles using silver nitrate as the precursor. The phenolic content in the extracts was assumed to be capable of reducing the metallic ions and also provide antimicrobial properties. The extracts contained water, alcohols, phenols, and amines based on FTIR spectra (Fig. 6.6), which created the ability for the nanoparticles to be coated with the organic compounds. Nanoparticles obtained had size between 22 and 40 nm and showed about 73–96% scavenging activity, higher than ascorbic acid, and had both antibacterial and antifungal activity. Aqueous extracts of areca leaves were used as a fuel to obtain ZnO nanoparticles by heating $Zn(NO_3)_2.6H_2O$ at 400 °C. Nanoparticles of about 29 nm were formed and had excellent antidiabetic

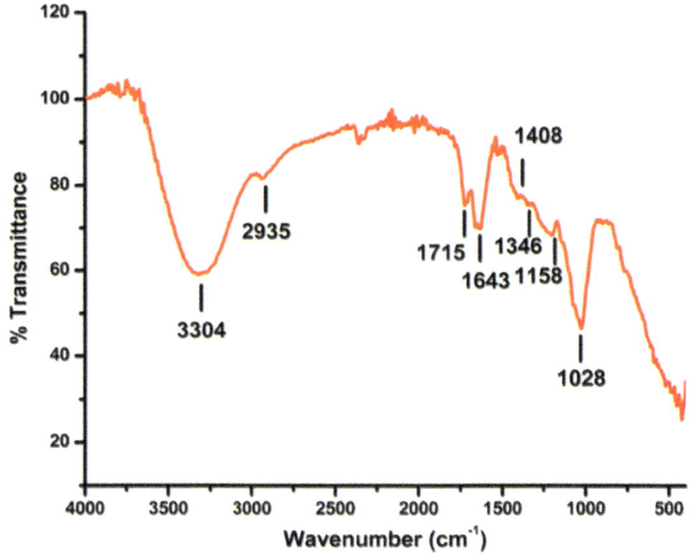

Fig. 6.6 FTIR spectrum of the components extracted from areca leaves (Shwetha 2021)

and anticancer activity (Shwetha 2020). Similarly, it was demonstrated that areca leaf extracts obtained by boiling the leaves in water at 60 °C for 30 min could be used to synthesize NiO nanoparticles using nickel nitrate hexahydrate $Ni(NO_3)_2 6H_2O)$ as the precursor. Nanoparticles with average size of 5.6 nm and having antidiabetic, anticancer activity against human lung cancer cells were obtained (Shwetha 2021).

6.3 Compost from Areca Leaves

The large availability of areca leaves at the farm level, and the composition of the leaves, make it a suitable source as manure or compost. However, the leaves have to be treated with different microbes to ensure sufficient availability of carbon and nitrogen for being considered as compost. Areca leaves were found to have higher C/N ratio compared to other parts of the plant, and when treated with microorganisms, the C/N increased to 60.5 from the initial level of 22.2 (Table 6.4), suggesting that the leaves could be preferred for preparing biocompost (Nagaraja 2014).

Table 6.4 Influence of microbial inoculants on carbon and nitrogen ratio (C:N) in arecanut leaf at periodical intervals (Nagaraja 2014)

Treatments	C:N ratio					
	30d	60d	90d	120d	150d	180d
T1	90.07	81.34	76.03	67.21	60.58	60.58
T2	83.80	72.95	68.67	62.18	49.38	48.17
T3	65.63	60.80	55.72	47.97	25.00	23.20
T4	58.33	53.37	48.04	40.69	22.95	22.24
T5	58.15	51.84	45.15	40.91	25.25	24.52
T6	63.82	54.14	48.06	44.84	30.48	29.26
T7	56.25	47.02	45.05	39.39	24.28	23.38
T8	71.20	63.27	56.25	48.53	34.94	34.17
T9	67.80	62.20	57.94	50.29	31.56	30.87
T10	67.59	62.35	60.59	54.26	31.74	30.10

6.4 Fibers from Areca Leaves

Although areca husks are mostly used as a source for fibers, the leaves are also lignocellulosic and highly fibrous. Hence, a few studies have been done on using areca leaves as a source for natural cellulose fibers. Attempts have also been made to extract natural fibers from the peduncle of areca leaves. Fibers extracted from the peduncle had 50% cellulose, high crystallinity index of 49%, and tensile strength between 88 and 184 MPa, similar to natural cellulose fibers obtained from other parts of the areca plants (Fig. 6.7). The rough surface of the fibers was suggested to provide good compatibility with the matrix leading to composites with good properties, which leads to promoting the use of a sustainable and renewable resource that is currently being disposed of as waste (Binoj 2024). Fibers extracted from areca sheath were reported to have 65% cellulose, 8% hemicellulose, and 18.6% lignin with crystallinity of up to 81%. Unlike previous reports, the fibers were found to be relatively weaker with a strength of 59 MPa and density of 0.9 g/cm^3. The fibers were thermally stable with residual mass of 33% even after heating to 500 °C (Suresh 2021). In another study, fibers extracted from areca sheath using 17.5% NaOH at 80 °C were found to have the properties required for use in natural fiber composites (Azmi 2023). Similarly, peduncle treated with 5% sodium hydroxide for 60 min enhanced strength of the fibers by 12.9% and crystallinity index by 14.2%. Fibers obtained were suggested to be suitable for developing biocomposites and could be considered as replacement for synthetic polymer-based fibers (Mansingh 2024).

Areca leaves, specifically the central part of the leaves, were separated (Fig. 6.8) and used for fiber extraction by immersing in water for 2 weeks and later beating the fibers with a wooden mallet. Alternatively, the leaf stalk was treated with various alkali concentrations (5 to 15%) at room temperature for 30 min. The fibers formed were washed thoroughly after treatment and used for characterization and composite

Fig. 6.7 Digital images of the penduncle in the areca nut tree, separated peduncle and fibers extracted (Binoj 2024). Reproduced with permission from Springer Nature

development. Density of the fibers obtained was between 1.09 and 1.34 g/cm^3, and average diameter was between 210 and 330 μm. Alkali treatment resulted in fibers with nearly 10% higher cellulose, 12% lower hemicellulose, and 2–3% lower lignin content compared to the fibers obtained through water retting. The results were also reflected in the considerably higher strength (486 MPa) and modulus (9.9 GPa) compared to the untreated and those fibers obtained after treating with 10 and 15% alkali. Fibers obtained were considered suitable for making composites and other applications (Shanmugasundaram 2018).

Alternatively, fibers obtained from the areca leaf stalks were combined with isophthalic resin and made into composites with 70% fibers and 30% matrix (Shanmugasundaram 2018). Both tensile and flexural strength and modulus of the composites reinforced with 15% alkali treated fibers were higher compared to the untreated and fibers treated with lower concentration of alkali due to the better interaction between the matrix and reinforcement (Hariharashayee 2020). Areca leaf stalks were treated by immersing them in water for 2 weeks and subjecting the material to mechanical beating. Later, the stalks were treated with alkali of three different concentrations (5, 10, and 15%) for 30 min. Fibers obtained were of different densities, and average

Fig. 6.8 Steps in preparing and extracting natural cellulose fibers from the areca leaves (Shanmugasundaram 2018). Reproduced with permission from Elsevier

Table 6.5 Properties of untreated and alkali-treated fibers obtained from areca leaf stalks (Shanmugasundram 2018a). Reproduced with permission through Elsevier

Type of fiber	Tensile strength, MPa	Tensile modulus, GPa	Strain at break, %
Untreated	335 ± 21	7.6 ± 1.1	4.4 ± 1.2
5% treated	486 ± 36	9.9 ± 1.5	4.9 ± 1.8
10% treated	424 ± 28	8.8 ± 1.3	4.8 ± 1.6
15% treated	396 ± 24	8.4 ± 1.3	4.7 ± 1.4

diameter was between 1.09 and 1.34 g/cm^3 and 210 and 330 μm. In terms of composition, the fibers had cellulose content between 57 and 63%, lignin between 7 and 4.5%, and moisture between was 9 and 1%. The treatment resulted in fibers having varying tensile properties, which were similar to many other fibers obtained from other biomass based materials (Shanmugasundaram 2018a) (Table 6.5).

6.5 Areca Trunk and Wood

Areca tree, typically found in tropical and subtropical climates, grows to an average height of 15 m, and the trunk is about 16 cm in diameter. Average lifespan of an areca tree is about 60 years. However, many trees do get damaged due to climatic conditions, diseases, etc. Trees are also felled regularly for changes in farming conditions and other requirements. Areca tree stems are excellent biomass and are generally used as wood for construction, furniture, and other applications. Areca trunks have also been used for ornamental applications. Properties of areca wood vary considerably depending on the source, variety, age, climatic conditions, and also on the location on the trunk (Chao 2020). Abandoning or discarding felled areca tree trunks has been considered to be an environmental burden. Various applications, composites, furniture, ornaments, carpets, etc., have been developed using areca tree trunks (Chao 2018).

In an attempt to utilize areca trees as an alternative to plywood, the stem of the trees were cut into 25 cm long pieces and passed through a chipper, and particles passing through the 4 and 20 mesh screens were collected. The particles obtained had an average length and thickness of 15 and 1.2 mm, respectively. These particles were combined with urea formaldehyde with or without the addition of wax (0.5, 1, or 1.5%). Panels were formed by compression molding at 150 °C for 10 min at a pressure of 5 MPa. Properties of the panels were dependent on the resin and wax content. Modulus of elasticity varied between 1310 and 2180 N/mm^2, modulus of resistance was between 10.5 and 14.8 N/mm^2, internal bond strength was between 0.52 and 0.78 N/mm^2 and screw holding capacity was between 390 and 590 N. These properties were considered to be similar or better than that of commercially available plywoods, and hence areca wood was suggested to be suitable for manufacturing furniture and other products (Lin 2008). Natural cellulose fibers were extracted from the areca stems and used as reinforcement for laminated polyester composites. Fibers were added to the resin in various orientations and with an overall fiber content of 15%. Properties of the composites were better than that of areca husk, bamboo, coir, or banana reinforced composites.

6.6 Areca Heartwood

The trunk of arecanut trees contains an outer sheath and an inner heartwood. The trees, when rendered aged or succumb to environmental factors, are considered as waste and disposed. In an effort to utilize the trunk of the areca tree, the heartwood, which was soft and spongy was separated by hand and cut into 1–2 cm length pieces and later powdered and sieved to obtain particles of 200 μm. These particles had a surface area of 7.3 m^2/g, pore size of 2.5 Å and pore volume of 4.7 cm^3/g. Ability of the particles to sorb cadmium (II) ions from aqueous solution was studied at various sorption parameters. The areca heartwood particles could attain equilibrium sorption in 30 min at pH 6.0. The biosorption process followed pseudo-second order rate reactions with maximum sorption of 10.7 mg/g, which was comparable to other biomass-based sorbents (Chakravarty 2010).

6.7 Conclusions

Areca trunk and leaves account for a major byproducts based on their weight but their availability is limited compared to the husks and sheaths. However, areca trees are cut, and leaves are pruned regularly and hence, they are available as byproducts. The leaves contain valuable compounds that show antimicrobial properties and could also be used for anticorrosive applications and for synthesis of nanoparticles. Some studies have suggested the use of the leaves and trunk for production of paper and as a source for ethanol.

Chapter 7
Future Perspectives

Ensuring the availability of affordable food, fiber, fuel, and materials for a rapidly growing global population will be a major challenge in the coming decades. Limited natural resources, rising labor costs, climate change, and geopolitical instability are key factors contributing to this challenge. Governments and research institutions worldwide are actively working to mitigate these impacts through sustainable strategies. Among the most effective approaches are maximizing the utilization of natural resources, adopting green processing technologies, and reducing reliance on non-renewable materials. In this context, the utilization of crop residues represents a highly promising solution. As a widely cultivated crop, areca generates substantial amounts of byproducts with unique structural and physicochemical properties. While areca husks, leaves, and sheaths have been explored for applications in fiber production, fuel generation, composites, and environmental remediation, there remains significant potential to develop novel applications and establish areca byproducts as renewable, sustainable, and cost-effective resources.

One of the most promising strategies for utilizing areca byproducts is their development as alternatives to plastic-based products, energy materials, and electromechanical components. For example, areca-derived materials show potential for applications in 3D printing, electrochemical energy storage, biosensing, and related technologies. Material functionalization can further enhance performance and expand application possibilities. However, the lignocellulosic composition and non-thermoplastic nature of areca biomass present processing challenges. These limitations can potentially be addressed through green chemical modification strategies, such as Diels–Alder dynamic covalent interactions between cellulose and lignin, enabling the development of advanced biobased materials. Through rational material design, biomass-derived conductive elastomers with tunable microporous architecture, high stretchability, resilience, and electrical responsiveness can be developed. Furthermore, the integration of automation, artificial intelligence (AI), and machine learning (ML) can improve biomass conversion processes for biofuel production, anaerobic digestion, and waste-to-energy systems. Areca byproducts

N. Reddy and P. Aramwit, *Areca catechu and its byproducts*,
https://doi.org/10.1007/978-3-032-22134-6_7

also hold promise for use in energy storage devices, catalysis, construction insulation materials, and environmental remediation applications, including carbon sequestration. With continued research and technological advancements, it is reasonable to anticipate a substantial increase in the utilization of areca byproducts, positioning them as valuable yet currently underutilized renewable and sustainable resources.

References

Abbas, Ghulam, Sabira Naqvi, Shaista Erum, Shakil Ahmed, and Ahsana Dar. Potential antidepressant activity of Areca catechu nut via elevation of serotonin and noradrenaline in the hippocampus of rats. *Phytotherapy Research* 27 (1) (2013): 39-45.

Adawiyah, Robiatul, Yulizah Hanifah, and Neza Rahayu Palapa. Analysis Congo red dye adsorption on hydrochar and nanohydrochar from Areca catechu: Synthesis, characterization, and comparative study. *Journal of Ecological Engineering* 26 (5) (2025).

Adnane, Imane, Hamza Taoumi, Karim Elouahabi, Khadija Lahrech, and Abdellah Oulmekki. Valorization of crop residues and animal wastes: Anaerobic co-digestion technology. *Heliyon* 10 (5) (2024).

Aisabor, W., Y. Lukman, A. J. Otaru, A. E. Anakhu, and S. K. Otoikhian. Optimisation and Kinetics of Rubber Seed Oil Biodiesel Production. Journal of Biofuels 7 (2) (2016): 53-61.

Akhter, Shirin, Maruf Hassan, Shahnaz Rahman, Afifa Sultana, Shejuty Shahreen, Joyonta Banik, Rownak Jahan, and Mohammed Rahmatullah. Anti-hyperglycemic activity studies on leaves and stems of Areca catechu L.(Arecaceae). Advances in Natural and Applied Sciences 8 (4) (2014): 233–236.

Akter, M., E. Jahan, and M. Hasan. Mechanical, thermal and morphological properties of pineapple and betel nut husk fiber reinforced hybrid polypropylene composites. In *IOP Conference Series: Materials Science and Engineering*, vol. 438, p. 012026. IOP Publishing, 2018.

Alshahrani, Hassan, and VR Arun Prakash. Effect of silane-grafted orange peel biochar and areca fibre on mechanical, thermal conductivity and dielectric properties of epoxy resin composites. *Biomass Conversion and Biorefinery* 14 (6) (2024): 8081–8089.

Anand, Praveena Bindiganavile, M. S. Aswathanarayan, Santhosh Nagaraja, A. Chandrashekar, H. D. Shivakumar, Sri Nugroho, and Muhammad Imam Ammarullah. Synthesis and Characterization of Mechanical Properties of Dammar Gum-Epoxy Bio-Composites with Areca Nut Husk and Banana Fiber as Reinforcements for Biomedical Applications. *Journal of Chemical Engineering of Japan* 57 (1) (2024): 2438111.

Anggini, Awanda Wira, Muhammad Adly Rahandi Lubis, Rita Kartika Sari, Antonios N. Papadopoulos, Petar Antov, Apri Heri Iswanto, Seng Hua Lee, Efri Mardawati, Lubos Kristak, and Ika Juliana. Cohesion and Adhesion Performance of Tannin-Glyoxal Adhesives at Different Formulations and Hardener Types for Bonding Particleboard Made of Areca (Areca catechu) Leaf Sheath. *Polymers* 15 (16) (2023): 3425.

Anik, Atique Ishrak, Md Moinul Islam, and Md Saiful Islam. Suitability of using areca nut fiber as reinforcing material in compressed stabilized earth blocks (CSEB) for low-cost housing. *Discover Civil Engineering* 1 (1) (2024): 17.

N. Reddy and P. Aramwit, *Areca catechu and its byproducts*,
https://doi.org/10.1007/978-3-032-22134-6

Anthikat, Reena R. Nelson, Antonysamy Micheal, and Savarimuthu Ignacimuthu. Aphrodisiac effect of Areca catechu L. and Pedalium murex in rats. *Journal of Men's Health* 10 (2) (2013): 65–70.

Anuar, Muhammad Fahmi, Yap Wing Fen, Muhammad Zakwan Azizan, Fida'I. Rahmat, Mohd Hafiz Mohd Zaid, Rahayu Emilia Mohamed Khaidir, and Nur Alia Sheh Omar. Sustainable production of arecanut husk ash as potential silica replacement for synthesis of silicate-based glass-ceramics materials. Materials 14 (5) (2021): 1141.

Aragonés, Marta Macias, and Fátima Arroyo Torralvo. How to select the best approach for circular economy assessment? 3D positioning framework, decision support tool and critical analysis for bio-based systems. *Environmental Impact Assessment Review* 106 (2024): 107493.

Ashok, R. B., C. V. Srinivasa, and B. Basavaraju. A review on the mechanical properties of areca fiber reinforced composites. *Science and Technology of Materials* 30 (2) (2018): 120-130.

Ashok, R. B., C. V. Srinivasa, and B. Basavaraju. Study on morphology and mechanical behavior of areca leaf sheath reinforced epoxy composites. Advanced Composites and Hybrid Materials 3 (2020): 365-374.

Attihalli Venkatachalaiah, Madhushri, Raju Nama Vasudeva Setty, Jobish Johns, Gajanan V. Honnavar, and M. V. Murugendrappa. Impact of Compatibilizer on the Electrical and Thermal Characteristics of Areca Fibre-Filled Polypropylene Composites. *Fibers and Polymers* (2024): 1–14.

Azmi, Sarah Adilah, Gusti Umindya Nur Tajalla, Muhammad Luthfi Ramadhani, Ade Wahyu Yusariarta Putra Parmita, and Arie Wibowo. Study of effect time alkalization on extraction of cellulose of areca leaf sheath fiber (Areca catechu). In *AIP Conference Proceedings*, vol. 2592 (1). AIP Publishing, 2023.

Babu, T. Narendiranath, Tanay Kucloury a, Mohit Kumar Jain, and R. V. Mangalaraja. Mechanical behaviour of areca nut, sunn hemp natural fibers and E-glass fibers reinforced with epoxy composites. *Engineering and advanced tehnology* 8 (6) (2019): 2249–8958.

Baidya, Kamal Sukla, and Upendra Kumar. Adsorption of cationic dye in aqueous solution by chemically modified Areca catechu husk. Desalination and Water Treatment 205 (2020): 422–434.

Baidya, Kamal Sukla, and Upendra Kumar. Adsorption of brilliant green dye from aqueous solution onto chemically modified areca nut husk. South African Journal of Chemical Engineering 35 (2021): 33–43.

Balguri, Praveen Kumar, A. Latha, Lakhvinder Kaur, Rajesh Verma, D. Sendil Kumar, S. Ramasree, Angajala Rama Krishna, Manzoore Elahi M. Soudagar, and N. Nagabhooshanam. Characterization of vinyl silane–treated areca nut woven fiber and bronze filler toughened polyester composite. *Biomass Conversion and Biorefinery* (2024): 1–11.

Banagar, Ashok R., Srinivasa Chikkol Venkateshappa, Sakshi Shantharam Kamath, and Basavaraju Bennehalli. Tensile and flexural properties of areca sheath fibers. Materials Today: Proceedings 5 (14) (2018): 28080–28088.

Banagar, Ashok, Srinivasa Venkateshappa Chikkol, and Basavaraju Bennehalli. Studies on physical and mechanical properties of untreated (raw) and treated areca leaf sheaths. *Materials Research Innovations* 25, no. 7 (2021): 404–411.

Banagar, Ashok R., GR Raghu Patel, and C. V. Srinivasa. Areca sheath fibers, their composites and applications. In *Plant Fibers, their Composites, and Applications*, pp. 75–110. Woodhead Publishing, 2022.

Banagar, Ashok R., GR Raghu Patel, C. V. Srinivasa, S. Dhanalakshmi, Hemaraju, and B. T. Ramesh. Extraction and characterization of hemicellulose and lignin contents of areca fiber. *Journal of the Indian Academy of Wood Science* 21 (1) (2024): 112–122.

Bardhan, Mondira, Tamanna Mamun Novera, Mumtahina Tabassum, Md Azharul Islam, Ali H. Jawad, and Md Atikul Islam. Adsorption of methylene blue onto betel nut husk-based activated carbon prepared by sodium hydroxide activation process. Water Science and Technology 82 (9) (2020): 1932–1949.

Baskar, A., and PS Syed Shabudeen. Factorial design of experiment model enables to optimize the variables in wastewater decolorization process by using areca husk activated carbon fibre. J Chem Pharm Res 7 (2015): 1500–1511.

Basker, A., PS Syed Shabudeen, and A. P. Shekhar. Adsorption Efficiency of Crystal Violet Onto Green House Gas Emitting Agro-Waste Obtained from Areca Husk Carbon. Asian Journal of Chemistry 26 (10) (2014a).

Basker, A., PS Syed Shabudeen, S. Daniel, and P. Vignesh Kumar. Adsorptive removal of malachite green from aqueous solution using areca husk carbon. Rasayan journal of chemistry 7 (1) (2014b): 1–5.

Basker, A., PS Syed Shabudeen, and D. Kalaiselvi. Adsorption of Cationic Dye onto Activated Carbon from Areca Husk Fibre and Optimization of Operating Conditions Using Design of Experiment. International Journal of Advanced Research in Bio-Technology, 3(1), 2015, 06–24.

Basnet, Prabin, Pawan Kumar Ojha, Deepak Gyawali, Kedar Nath Ghimire, and Hari Paudyal. Thermochemical study of Cr (VI) sequestration onto chemically modified Areca catechu and its recovery by desorptive precipitation method. *Heliyon* 8 (8) (2022).

Basnet, Prabin, Deepak Gyawali, Kedar Nath Ghimire, and Hari Paudyal. Phosphoric acid activated Areca catechu as an efficient biosorbent for Cr (VI) sequestration from simulated and industrial effluent. *Separation Science and Technology* (2025): 1–22.

Basrur, Deepa, and J. Ishwara Bhat. An investigation on the characterization of activated carbon from areca leaves and their adsorption nature towards different dyes. Global Nest Journal 21 (2019): 124–130.

Ben Samuel, J., S. Julyes Jaisingh, R. Ramadoss, and T. Maridurai. Mechanical, wear and fatigue behaviour of neem oil and nanosilica toughened areca fibre reinforced epoxy resin composite. *Silicon* 15 (8) (2023): 3525–3533.

Benegal, Vivek, Ravi P. Rajkumar, and Kesavan Muralidharan. Does areca nut use lead to dependence. *Drug and alcohol dependence* 97, no. 1-2 (2008): 114-121.

Bera, Sounak, and Kaustubha Mohanty. Areca nut (Areca catechu) husks and Luffa (Luffa cylindrica) sponge as microbial immobilization matrices for efficient phenol degradation. Journal of Water Process Engineering 33 (2020): 100999.

Bhagyamma, C. V., S. Sharad, Shriganesh G. Gudigar, and S. V. Reshma. Bioprocessing of Areca husk in solid state fermentation for cellulase production using Trichoderma viride. Indian Journal of Applied Research, 1(3), 2011: 16–17.

Bhandare, Amol, Ajay Kshirsagar, Neeraj Vyawahare, Pushpendra Sharma, and Reshma Mohite. "Evaluation of anti-migraine potential of Areca catechu to prevent nitroglycerin-induced delayed inflammation in rat meninges: Possible involvement of NOS inhibition. *Journal of ethnopharmacology* 136 (1) (2011): 267-270.

Bhandare, Amol M., Ajay D. Kshirsagar, Neeraj S. Vyawahare, Avinash A. Hadambar, and Vrushali S. Thorve. Potential analgesic, anti-inflammatory and antioxidant activities of hydroalcoholic extract of Areca catechu L. nut. *Food and Chemical toxicology* 48, no. 12 (2010): 3412–3417.

Bharat, Masuram, Deepak Kumar Verma, Venkatesh Shanbhag, Ravindra Singh Rajput, Deepak Nayak, and Arul Amuthan. Ethanolic extract of oral Areca catechu promotes burn wound healing in rats. *Int J Pharm Sci Rev Res* 25 (2) (2014): 145–148.

Bhat, Ravishankar, Sharanabasava Ganachari, Raghunandan Deshpande, Ggg Ravindra, and A. Venkataraman. Rapid biosynthesis of silver nanoparticles using areca nut (areca catechu) extract under microwave-assistance. *Journal of Cluster Science* 24 (2013): 107–114.

Bhat, Ravi, S. Sujatha, Bhavishya, U. K. Priya, Alka Gupta, and Anok Uchoi. Arecanut (Areca catechu L.). In *Soil Health Management for Plantation Crops: Recent Advances and New Paradigms*, pp. 177–206. Singapore: Springer Nature Singapore, 2024

Binoj, J. S., R. Edwin Raj, B. S. S. Daniel, and S. S. Saravanakumar. Optimization of short Indian Areca fruit husk fiber (Areca catechu L.)–reinforced polymer composites for maximizing mechanical properties. *International Journal of Polymer Analysis and Characterization* 21 (2), (2016a): 112–122.

Binoj, J. S., R. Edwin Raj, V. S. Sreenivasan, and G. Rexin Thusnavis. Morphological, physical, mechanical, chemical and thermal characterization of sustainable Indian areca fruit husk fibers (Areca catechu L.) as potential alternate for hazardous synthetic fibers. Journal of Bionic Engineering 13 (1) (2016b): 156–165.

Binoj, J. S., R. Edwin Raj, and N. Manikandan. Development of Bio-Composites From Industrial Discarded Fruit Fibers. In *Natural Polymers*, pp. 41–68. Apple Academic Press, 2022.

Binoj, Joseph Selvi, Mariatti Jaafar, Bright Brailson Mansingh, and Raman Raghu. Viscoelastic and heat-resistant behavior of surface-treated areca fruit husk fiber-reinforced polymer composites. *Iranian Polymer Journal* 32 (7) (2023): 841–853.

Binoj, Joseph Selvi, Mariatti Jaafar, Bright Brailson Mansingh, and Govindarajan Bharathiraja. Extraction and characterization of novel cellulosic biofiber from peduncle of Areca catechu L. biowaste for sustainable biocomposites. *Biomass Conversion and Biorefinery* 14 (17) (2024): 20359–20367.

Boopathi, C., N. Natarajan, R. K. Sanjeev, V. Meganathan, and P. Kavin. "Influence of red mud particles on thermo-mechanical behaviour of areca fiber reinforced polymer composites." In *AIP Conference Proceedings*, 2128 (1). AIP Publishing, 2019.

Borah, Jyotishmoy, and Nobarun Dutta. Development and properties evaluation of betel nut fibres composite material. *Materials Today: Proceedings* 5 (1) (2018): 2229-2233.

Boro, Simangsa, Bipul Das, Sujata Brahma, Bidangshri Basumatary, Siri Fung Basumatary, and Sanjay Basumatary. Biodiesel production using areca nut (Areca catechu L.) leaf ash-K2CO3 catalyst via transesterification from an oil blend of three different feedstocks. *Sustainable Chemistry for the Environment* 8 (2024): 100164.

Brailson Mansingh, Bright, Joseph Selvi Binoj, Amirthaiah Amala Mithin Minther Singh, and Antony Sagai Francis Britto. Influence of SiC nanoparticles on properties of alkali-treated areca fruit husk fiber/hybrid polymer composites. Journal of Applied Polymer Science 140 (10) (2023): e53591.

Byun, Sung-June, Hee-Sook Kim, Seon-Min Jeon, Yong Bok Park, and Myung-Sook Choi. Supplementation of Areca catechu L. extract alters triglyceride absorption and cholesterol metabolism in rats. *Annals of nutrition and metabolism* 45 (6) (2001): 279–284.

Cao, L-M., B. Zhang, H-Y. Luo, K. Zhou, Y-F. Yu, Z-Z. Li, G-R. Wang et al. Areca nut production, imports, and their impact on oral cancer Incidence. *Journal of dental research* (2025): 00220345251344895.

Chakravarty, Paresh, N. Sen Sarma, and H. P. Sarma. Biosorption of cadmium (II) from aqueous solution using heartwood powder of Areca catechu. Chemical Engineering Journal 162 (3) (2010): 949–955.

Chanakya, Hoysall N., and Malayil Sreesha. Anaerobic retting of banana and arecanut wastes in a plug flow digester for recovery of fiber, biogas and compost. Energy for sustainable development 16 (2) (2012): 231-235.

Chandra, Julie, Neena George, and Sunil K. Narayanankutty. Isolation and characterization of cellulose nanofibrils from arecanut husk fibre. Carbohydrate polymers 142 (2016): 158-166.

Chandra, Julie, Bipinbal PK, Renju VS, Vidya Raman, Bindu Sharmila TK, Sreesha Sasi, and Jolly V. Antony. Bionanocomposites based on natural rubber and cellulose nanofibrils from arecanut husk: rheological, mechanical and thermal characterizations. *Journal of Polymer Research* 29 (6), (2022): 217.

Chandrappa, Ravi Kumar, and Sakshi Shantharam Kamath. X-ray diffraction and thermo gravimetric analysis of surface modified areca sheath fibre epoxy composites. *Journal of Mechanical and Energy Engineering* 6 (2) (2022).

Chang, Yung-Fu, Ting-Yuan Liu, Shao-Tung Liu, and Chao-Neng Tseng. Arecoline inhibits myogenic differentiation of C2C12 myoblasts by reducing STAT3 phosphorylation. *Food and chemical toxicology* 50 (10) (2012): 3433–3439.

Chang, Lien-Yu, Hsiao-Ching Wan, Yu-Lin Lai, I-Chiang Chou, Yen-Ting Chen, and Shan-Ling Hung. Areca nut extracts increased the expression of cyclooxygenase-2, prostaglandin E2 and interleukin-1α in human immune cells via oxidative stress. *Archives of Oral Biology* 58, no. 10 (2013): 1523–1531.

Chang, Yung-Fu, Ting-Yuan Liu, and Shao-Tung Liu. Arecoline inhibits and destabilizes agrin-induced acetylcholine receptor cluster formation in C2C12 myotubes. *Food and chemical toxicology* 60 (2013): 391–396.

Chao, F. L., C. K. Wu, and A. K. Chao. Demonstration on areca catechu tree reuse with supporting of information technology. In *Journal of Physics: Conference Series*, 989 (1), p. 012010. IOP Publishing, 2018.

Chao, Fang-Lin, Te-Hsin Yang, and Je-Yau Wu. Material Reuse Design of Areca Catechu Tree. https://ir.lib.cyut.edu.tw/bitstream/310901800/37235/2/B1-Wood.edited+Ho.pdf

Chavan, Yogita V., and Rekha S. Singhal. Separation of polyphenols and arecoline from areca nut (Areca catechu L.) by solvent extraction, its antioxidant activity, and identification of polyphenols. *Journal of the Science of Food and Agriculture* 93 (1)0 (2013): 2580–2589.

Chen, Da-Juan, Jacob B. Landis, Hong-Xin Wang, Qing-Hui Sun, Qiao Wang, and Hua-Feng Wang. "Plastome structure, phylogenomic analyses and molecular dating of Arecaceae." *Frontiers in Plant Science* 13 (2022): 960588.

Chethan, G., K. C. Sunil, Mahagundappa R. Maddani, and Y. Narayana. A Study on Shore D Hardness of Areca Husk Fibre Reinforced Polyester Resin Composite: Impact of Fibre Maturity. In *Materials Science Forum*, vol. 1111, pp. 83–88. Trans Tech Publications Ltd, 2023.

Chethan, M. R., SG Gopala Krishna, R. Chennakeshava, and D. Mahesh. Study on tensile analysis of untreated chopped natural areca sheath fiber reinforced polymer matrix bio-composites. International Journal of Engineering Research and Advancement Technology 2 (2016): 338–393.

Chethan, M. R., Parvatini Sri Naga Venkat, GS Gopala Krishna, R. Chennakesava, and P. J. M. T. P. Vijay. Dynamic vibrational analysis on areca sheath fibre reinforced bio composites by fast fourier analysis. Materials Today: Proceedings 5 (9) (2018): 19330–19339.

Chinnabhandar, Raghuram K., Doma Hemanth Kumar, R. G. Yatish, and A. U. Ravi Shankar. Laboratory evaluation of use of areca fibres in SMA mixes." International Journal of Pavement Engineering 24 (2) (2023): 2076090.

Chou, Tzu-min, and Jin-Long Hong. Areca nut–derived porous carbons for supercapacitor and CO2 capture applications. *Ionics* 26 (3) (2020): 1419-1429.

Choudhary, Purushottam Kumar, Priyabrat Pradhan, and Alok Satapathy. Study on mechanical and tribological characteristics of epoxy composites reinforced with short areca-catechu fibers. *Journal of Natural Fibers* 19 (13) (2022): 6602–6614.

Chowdari, G. Kishore, and DVV Krishna Prasad. Mechanical Characterization of Areca Fiber and Coconut Shell Powder Reinforced Hybrid Composites. In *Materials Science Forum*, vol. 1034, pp. 61–71. Trans Tech Publications Ltd, 2021.

Chungcharoen, Thatchapol, and Naruebodee Srisang. Preparation and characterization of fuel briquettes made from dual agricultural waste: Cashew nut shells and areca nuts. Journal of Cleaner Production 256 (2020): 120434.

CS, Julie Chandra, Bipinbal PK, Renju VS, Vidya Raman, Bindu Sharmila TK, Sreesha Sasi, and Jolly V. Antony. Bionanocomposites based on natural rubber and cellulose nanofibrils from arecanut husk: Rheological, mechanical and thermal characterizations. Journal of Polymer Research 29 (6) (2022): 217.

Development And Characterization Of Low Thermal Conductivity Material From Reinforced Areca Husk Fiber Composite - Composites: Mechanics, Computations, Applications: An International Journal, 13 (3), 2022.

Daioglou, Vassilis, Elke Stehfest, Birka Wicke, Andre Faaij, and Detlef P. van Vuuren. Projections of the availability and cost of residues from agriculture and forestry. *Gcb Bioenergy* 8(2) (2016): 456–470.

Dangore-Khasbage, Suwarna, Rahul R. Bhowate, and Monika Khubchandani. Chemical composition of areca nut and its adverse effects on human health. *Cureus* 15 (8) (2023).

Dangore-Khasbage, Suwarna, Rahul R. Bhowate, Monika Khubchandani, and Rahul Bhowate. "Chemical composition of areca nut and its adverse effects on human health." *Cureus* 15, no. 8 (2023).

Das, Sekhar, Atin Chaudhuri, and Amiya Kumar Singha. Characterization of lignocellulosic fibres extracted from agricultural biomass: arecanut leaf sheath. The Journal of The Textile Institute (2020): 1–8.

Das, Sekhar, Atin Chaudhuri, and Amiya Kumar Singha. Fabrication, microstructural and mechanical properties of arecanut leaf sheath fibre reinforced polyester resin composites. The Journal of The Textile Institute 113 (9) (2022): 1942–1949.

Dash, Shib Shankar. Rapid biosynthesis of Areca nut seed extract mediated gold nanoparticles and its use in catalytic reduction of 4-nitrophenol to 4-aminophenol. *Asian J Pharm Pharmacol* 5 (2019): 895-900.

Dash, Shib Shankar, Jhimli Banerjee, Sovan Samanta, Biplab Giri, and Sandeep Kumar Dash. Microwave-assisted fabrication of silver nanoparticles utilizing seed extract of areca catechu with antioxidant potency and evaluation of antibacterial efficacy against multidrug resistant pathogenic bacterial strains. *BioNanoScience* 12 (1) (2022): 210–227.

Deepak, K. B., and N. A. Jnanesh. An experimental study of various characteristics of biomass briquettes made from areca leaves: An alternative source of energy. (2013): 3–08.

Deepak, K. B., B. J. Manujesh, Vivek, and B. K. Yashas. Development and study of fuel briquettes from areca leaves: A potential renewable energy source. In AIP Conference Proceedings, vol. 2080, no. 1, p. 030004. AIP Publishing LLC, 2019.

Deepak, K. B., and N. A. Jnanesh. Experimental analysis of physical and fuel characteristics of areca leaves briquette. Int. J. Sci. Res 4 (1) (2015): 1261-64.

Desai, Raghuveer H., L. Krishnamurthy, and T. N. Shridhar. Influence of physical modification on Indian originated areca fiber based polypropylene composites: mechanical and dielectric properties characterization. In *AIP Conference Proceedings*, 2057 (1), p. 020052. AIP Publishing LLC, 2019.

Deshmukh, P. S., P. G. Patil, P. U. Shahare, G. B. Bhanage, J. S. Dhekale, K. G. Dhande, and V. V. Aware. Effect of mechanical and chemical treatments of arecanut (areca catechu L.) fruit husk on husk and its fibre. Waste Management 95 (2019): 458–465.

Dhanalakshmi, Sampathkumar, Punyamoorthy Ramadevi, and Bennehalli Basavaraju. A study of the effect of chemical treatments on areca fiber reinforced polypropylene composite properties. *Science and Engineering of Composite Materials* 24 (4) (2017): 501-520.

Dhanalakshmi, S., P. Ramadevi, and B. Basavaraju. Areca fiber reinforced epoxy composites: Effect of chemical treatments on impact strength. *Oriental Journal of Chemistry* 31 (2) (2015a): 763–769.

Dhanalakshmi, S., P. Ramadevi, and B. Basavaraju. Influence of chemical treatments on flexural strength of areca fiber reinforced epoxy composites. *Chemical Science* 4 (2) (2015b): 409–418.

Dhanapal, Duraibabu, S. Ananda Kumar, and J. Ranjitha. Areca fiber/epoxy nano-composite reinforced with amine-functionalized nano-alumina-enhancement in thermal and thermo-mechanical properties. *Materials Today Communications* 33 (2022): 104798.

Dhilip, Jafrey Daniel James, Vijay Raghunathan, Ramesh Mohan, Vinod Ayyappan, Sanjay Mavinkere Rangappa, and Suchart Siengchin. Mechanical and flammability properties of ultrasonically processed silane-treated areca-banana fiber-reinforced epoxy composites for lightweight applications. *Biomass Conversion and Biorefinery* (2024): 1–14.

Dhilipkumar, Thulasidhas, Raja Venkatesan, Seong-Cheol Kim, Arun Prasad Murali, Karthik V. Shankar, and Tahani Mazyad Almutairi. Assessing the structural and free vibrational performance of areca/ramie fibre composite reinforced with graphene nanofiller. *Industrial Crops and Products* 222 (2024): 119599.

Dissanayake, D. G. K., Dakshitha Weerasinghe, T. D. R. Perera, M. M. A. L. Bandara, S. K. T. Thathsara, and S. Perera. A sustainable transparent packaging material from the arecanut leaf sheath. *Waste and Biomass Valorization* (2021): 1–18.

Divakar, L., Abhishek Pulgur Venkatesha Babu, R. K. Chethan Gowda, and S. Nithin Kumar. Experimental study on mechanical properties of areca nut fibre-reinforced self-compacting concrete. In *Green Buildings and Sustainable Engineering: Proceedings of GBSE 2019*, pp. 313–321. Singapore: Springer Singapore, 2020.

El Bourakadi, Khadija, Fatima-Zahra Semlali, Maryama Hammi, and Mounir El Achaby. A review on natural cellulose fiber applications: Empowering industry with sustainable solutions. *International Journal of Biological Macromolecules* 281 (2024): 135773.

Elanthikkal, Silviya, Tania Francis, and Sultan Akhtar. Utilization of areca nut leaf sheath fibers for the extraction of cellulose whiskers. *Journal of Natural Fibers* 18 (9) (2021): 1261-1273.

Fadli, Ahmad. Dissolving pulp from Areca catechu l. fruit husk fibers by water-pre-hydrolysis, soda-cooking, and chlorine-free bleaching processes. Materials Today: Proceedings 87 (2023): 311–314.

Ganesh, Saikrishnan, Yoganjaneyulu Gunda, Sudhan Raj Jegan Mohan, Vijay Raghunathan, and Jafrey Daniel James Dhilip. Influence of stacking sequence on the mechanical and water absorption characteristics of areca sheath-palm leaf sheath fibers reinforced epoxy composites. Journal of Natural Fibers 19 (5) (2022): 1670-1680.

Garg, Apurva, Pankaj Chaturvedi, and Prakash C. Gupta. A review of the systemic adverse effects of areca nut or betel nut. *Indian journal of medical and paediatric oncology* 35 (1) (2014): 3-9.

Garg, Krati, Anuj Kumar, Vidisha Kizhakkethil, Pramod Kumar, and Shalini Singh. Overlap in oncogenic and pro-inflammatory pathways associated with areca nut and nicotine exposure. *Cancer Pathogenesis and Therapy* 2 (3) (2024): 187-194.

Gautam, Anirudh Muralidharan, and Nídia Caetano. Study, design and analysis of sustainable alternatives to plastic takeaway cutlery and crockery. Energy Procedia 136 (2017): 507–512.

GC, Anjan Kumar, Yadav D. Bodke, B. Manjunatha, and O. Nagaraja. Novel Areca Nut extract mediated preparation of TiO2 nanoparticles and its catalytic synthesis of 3, 3′-(phenyl methylene) bis-(4-hydroxy-2H-chromen-2-one) derivatives. *Chemical Data Collections* 38 (2022): 100832.

Gireeshkumar, B. C., B. Sharath Kumar, Y. Arthoba Nayaka, H. B. Muralidhara, Krishna Venkatesh, and A. Ramadas. Improving energy storage efficiency through carbon doping of niobium oxide nanomaterials derived from areca husk in redox flow batteries and supercapacitors. Journal of Energy Storage 71 (2023): 108139.

Girish, C. R., and Geoju M. George. Phenol removal from wastewater Using arecanut husk (areca catechu) As adsorbent. International Journal of Mechanical Engineering and Technology 8 (12) (2017): 1–9.

Gogoi, Debajeet, Neonjyoti Bordoloi, Ritusmita Goswami, Rumi Narzari, Ruprekha Saikia, Debashis Sut, Lina Gogoi, and Rupam Kataki. Effect of torrefaction on yield and quality of pyrolytic products of arecanut husk: an agro-processing wastes. Bioresource technology 242 (2017): 36-44.

Gokarneshan, Narayanan, Venkatesan Sathya, Jayagopal Lavanya, Shaistha Shabnum, and Sona M. Anton. A review of significant advances in areca fiber composites. *Next-Generation Textiles* (2022). A Review of Significant Advances in Areca Fiber Composites | IntechOpen.

Gokul, P. V., Pratap Singh, Vishwajeet P. Singh, and Ashish N. Sawarkar. Thermal behavior and kinetics of pyrolysis of areca nut husk. *Energy Sources, Part A: Recovery, Utilization, and Environmental Effects* 41 (23) (2019): 2906–2916.

Gour, Khalid Ahmed, Ravi Ramadoss, and Thirumalini Selvaraj. Revamping the traditional air lime mortar using the natural polymer–Areca nut for restoration application. *Construction and Building Materials* 164 (2018): 255–264.

Gudapati, SP Kumar, K. Supraja, and AV Ratna Prasadc. Thermal Conductivity Characterization of Areca Palm Fiber Reinforced Polymer Composites. International Journal of pure and applied mathematics 119 (17) (2018): 1871–1880.

Gujre, Nihal, Sudip Mitra, Richa Agnihotri, Mahaveer P. Sharma, and Debaditya Gupta. Novel agrotechnological intervention for soil amendment through areca nut husk biochar in conjunction with vetiver grass. Chemosphere 287 (2022): 132443.

Guna, Vijaykumar, Manikandan Ilangovan, K. C. V. A. K. Adithya, Akshay Koushik CV, C. V. Srinivas, S. Yogesh, G. S. Nagananda, Krishna Venkatesh, and Narendra Reddy. Biofibers and biocomposites from sabai grass: A unique renewable resource. *Carbohydrate Polymers* 218 (2019): 243–249.

Guo, Ziyuan, Zihan Wang, Yinghua Luo, Lingjun Ma, Xiaosong Hu, Fang Chen, Daotong Li, and Min Jia. Extraction and identification of bioactive compounds from areca nut (Areca catechu L.) and potential for future applications. *Food Frontiers* 5 (5) (2024): 1909–1932.

Gupta, Alpana K., Sonam Tulsyan, Nisha Thakur, Vishwas Sharma, Dhirendra N. Sinha, and Ravi Mehrotra. Chemistry, metabolism and pharmacology of carcinogenic alkaloids present in areca nut and factors affecting their concentration. *Regulatory Toxicology and Pharmacology* 110 (2020): 104548.

Gupta, Shaswat Vikram, V. V. Kulkarni, and Md Ahmaruzzaman. Fabrication of a bio-adsorbent material by grafting CeO2 quantum dots (QDts) over Areca nut shell biochar using Saccharum officinarum extract as a solvent/capping agent for adsorption of Methylene blue dye: Synthesis, material analyses, adsorption kinetics and isotherms studies. *Colloids and Surfaces A: Physicochemical and Engineering Aspects* 680 (2024): 132611.

Habeeba, Umme, and Narasimha Raghavendra. Green synthesis and antioxidant potency of silver nanoparticles using arecanut seed extract. Nano-Structures & Nano-Objects 38 (2024): 101118.

Hariharan, Gajendiran, Pattipati Chinna Krishnamachary, Joseph Selvi Binoj, and Bright Brailson Mansingh. Influence of SiC nanoparticles reinforcement in areca/tamarind hybrid biopolymer composites: thermo-mechanical, tribological and morphological features. *Journal of Bionic Engineering* 20 (4) (2023): 1723–1736.

Hariharashayee, D., M. P. Prabakaran, C. Chelladurai, S. Manikandan, N. S. Mohan, and M. Pradeepa. Experimental investigation of raw and alkali treated areca leaf stalk and sisal fibres reinforced hybrid polymer composites. Wutan Huatan Jisuan Jishu 16 (6) (2020): 273-283.

Hasan, Noor Md Sadiqul, Nur Mohammad Nazmus Shaurdho, Md Habibur Rahman Sobuz, Md Montaseer Meraz, Md Abdul Basit, Suvash Chandra Paul, and Md Jihad Miah. Rheological, mechanical, and micro-structural property assessment of eco-friendly concrete reinforced with waste areca nut husk fiber. *Sustainability* 15 (19) (2023): 14131.

Hassan, M. Masudul, Manfred H. Wagner, H. U. Zaman, and Mubarak A. Khan. Physico-mechanical performance of hybrid betel nut (Areca catechu) short fiber/seaweed polypropylene composite. *Journal of Natural Fibers* 7 (3) (2010): 165–177.

Hegde, Rajeev V., Arnab Ghosh, Arvind H. Jadhav, Aatika Nizam, Siddappa A. Patil, Febina Peter, and Ramesh B. Dateer. Biogenic synthesis of Pd-nanoparticles using Areca Nut Husk Extract: A greener approach to access α-keto imides and stilbenes. New Journal of Chemistry 45 (35) (2021): 16213-16222.

Hegde, Gautam, H. T. Bhavana, Anantha Sunil Maligi, M. Vandana, Gurumurthy Hegde, and Gopalkrishna Hegde. Enhancement of tensile strength and elastic modulus using bio-waste based carbon nanospheres doped polymer nanocomposites. *Journal of Polymer Research* 31 (6), (2024): 157.

Herlinawati, H., M. Yusuf, M. I. Siregar, and M. Yunus. Mechanical properties of polyethylene matrix composites with areca fruit peel powder filler (Areca catechu) and sugar cane fiber (Saccharum officinarum L). In *Journal of Physics: Conference Series*, vol. 2193 (1), p. 012058. IOP Publishing, 2022.

Ho, W-H., Y-Y. Lee, L-Y. Chang, Y-T. Chen, T-Y. Liu, and S-L. Hung. Effects of areca nut extract on the apoptosis pathways in human neutrophils. Journal of periodontal research 45 (3), (2010): 412–420.

Holland, Claire, and Philip Shapira. Building the bioeconomy: A targeted assessment approach to identifying biobased technologies, challenges and opportunities. *Engineering Biology* 8 (1) (2024): 1-15.

Hu, Xiaoping, Lingfeng Wu, Minjie Du, and Lu Wang. Eco-friendly synthesis of size-controlled silver nanoparticles by using Areca catechu nut aqueous extract and investigation of their potent antioxidant and anti-bacterial activities. Arabian Journal of Chemistry 15 (5) (2022): 103763.

Huang, Pei-Ling, Chin-Wen Chi, and Tsung-Yun Liu. Areca nut procyanidins ameliorate streptozocin-induced hyperglycemia by regulating gluconeogenesis. *Food and chemical toxicology* 55 (2013): 137-143.

Huang, Pei-Ling, Chin-Wen Chi, and Tsung-Yun Liu. Effects of Areca catechu L. containing procyanidins on cyclooxygenase-2 expression in vitro and in vivo. *Food and Chemical Toxicology* 48 (1) (2010): 306–313.

Huang, Gang, Deyong Zeng, Tisong Liang, Yaping Liu, Fang Cui, Haitian Zhao, and Weihong Lu. Recent advance on biological activity and toxicity of arecoline in edible areca (betel) nut: a review. *Foods* 13 (23) (2024): 3825.

Hugar, Priyanka, Ankita Dutta, S. Srilakshmi, Prasanna D. Belur, Keyur Raval, and Regupathi Iyyaswami. Phenolic profile of unripe areca nuts cultivated in various districts of Karnataka, India. *JSFA Reports* 4 (2) (2024): 102–113.

Islam, Md Zahidul, ANM Masudur Rahman, and Hosne Ara Begum. Comparative study of areca fiber reinforced polyester composites and plywood. *Eur Sci J October* (2018).

Jagadeesh, Praveenkumara, Madhu Puttegowda, Yashas Gowda Thyavihalli Girijappa, Sanjay Mavinkere Rangappa, and Suchart Siengchin. Carbon fiber reinforced areca/sisal hybrid composites for railway interior applications: Mechanical and morphological properties. *Polymer Composites* 43 (1) (2022): 160–172.

Jain, Vipin, Apurva Garg, Mark Parascandola, Pankaj Chaturvedi, Samir S. Khariwala, and Irina Stepanov. Analysis of alkaloids in areca nut-containing products by liquid chromatography–tandem mass spectrometry. *Journal of agricultural and food chemistry* 65 (9) (2017): 1977-1983.

Jam, Neda, Reza Hajimohammadi, Parvin Gharbani, and Ali Mehrizad. Evaluation of antibacterial activity of aqueous, ethanolic and methanolic extracts of areca nut fruit on selected bacteria. *BioMed Research International* 2021 (1) (2021): 6663399.

Jaroenkietkajorn, Ukrit, Shabbir H. Gheewala, Rattanawan Mungkung, Napat Jakrawatana, Thapat Silalertruksa, Naruetep Lecksiwilai, Jittima Prasara-A, and Pariyapat Nilsalab. Challenges and opportunities of bio-circular-green economy for agriculture. *Circular economy and sustainability* 4 (3) (2024): 1729-1750.

Jayaprakash, Kuzhandaivel, Atieh Osama, Rajinikanth Rajagopal, Bernard Goyette, and Obulisamy Parthiba Karthikeyan. Agriculture waste biomass repurposed into natural fibers: a circular bioeconomy perspective. *Bioengineering* 9 (7) (2022): 296.

Jeng, J. H., M. C. Chang, and L. J. Hahn. Role of areca nut in betel quid-associated chemical carcinogenesis: current awareness and future perspectives. *Oral oncology* 37 (6) (2001): 477-492.

Jose, Jasmine, C. J. Binish, Jobish Johns, Sony J. Chundattu, and A. V. Vijayasankar. Agricultural waste valorisation–Novel Areca catechu L. residue blended with PVA-Chitosan for removal of chromium (VI) from water–Characterization, kinetics, and isotherm studies. *Industrial Crops and Products* 222 (2024): 120013.

Joshi, Rahul, Pramendra Kumar Bajpai, and Samrat Mukhopadhyay. Effect of reinforcing waste bagasse ash and eggshells as fillers on mechanical and viscoelastic characteristics of areca/epoxy hybrid composites. Biomass Conversion and Biorefinery (2024): 1–21.

Joshi, Sahira, Rekha Goswami Shrestha, Raja Ram Pradhananga, Katsuhiko Ariga, and Lok Kumar Shrestha. High surface area nanoporous activated carbons materials from Areca catechu nut with excellent iodine and methylene blue adsorption. *C* 8 (1) (2021): 2.

Jothibasu, S., S. Mohanamurugan, R. Vijay, D. Lenin Singaravelu, A. Vinod, and M. R. Sanjay. Investigation on the mechanical behavior of areca sheath fibers/jute fibers/glass fabrics reinforced hybrid composite for light weight applications."*Journal of Industrial Textiles* 49 (8) (2020): 1036–1060.

KALASEE, Wachara, Visit EAKVANICH, Natworapol RACHSIRIWATCHARABUL, Wassachol WATTANA, Panya DANGWILAILUX, and Putipong LAKACHAIWORAKUN. Sound Absorption Properties of Natural Fiber Composite from Areca Nut Shells Fibers with Polyvinyl Alcohol. *Journal of the Korean Wood Science and Technology* 53 (2) (2025): 105–118.

Kalita, Pankaj, U. S. Dixit, P. Mahanta, and U. K. Saha. A novel energy efficient machine for plate manufacturing from areca palm leaf sheath. (2008).

Kamath, Sakshi Shantharam, and Basavaraju Bennehalli. Potential of using areca fibres in composite fabrication. Materials Today: Proceedings 44 (2021): 4143–4149.

Kamath, Sakshi Shantharam, and Basavaraju Bennehalli. Surface Modification of Areca Fibre by Benzoyl Peroxide and Mechanical Behaviour of Areca-Epoxy Composites. Material Science Research India 18 (1) (2021): 48–55.

Kamath, Sakshi S., D. N. Punith, S. Preetham, S. N. Gautham, Janardhan, K. Lalith Yashwanth, and Basavaraju Bennehalli. Tensile and flexural behaviour of areca husk fibre reinforced epoxy composite. In Advances in Metrology and Measurement of Engineering Surfaces: Select Proceedings of ICFMMP 2019, pp. 35–43. Springer Singapore, 2021.

Kamath, Sakshi Shantharam, Basavaraju Bennehalli, and Azeem Pasha. Surface modification of areca fibre by benzoyl peroxide and mechanical behaviour of areca-epoxy composites. *Mat. Sci. Res. India* 18 (1) (2021): 0–7.

Kavya, N., N. Nagaraja, R. B. Ashok, and S. Latha. "Determination of thermal insulating properties of cenosphere powder-filled areca husk fibre reinforced epoxy composites." *Journal of the Indian Academy of Wood Science* 20 (2) (2023): 149-156.

Kencanawati, C. I. P. K., N. P. G. Suardana, I. K. G. Sugita, and I. W. B. Suyasa. Characteristics of Fiber Treatments on Tensile and Impact Strengths of Pine Resin/Areca Husk Fiber Biocomposites. *International Journal of Civil Engineering and Technology* 11 (1) (2020).

Khan, Mohammad KA, M. Faisal, and V. R. Arun Prakash. Effect of silane coupling grafted polyethylene terephthalate foam and areca fruit fiber reinforced chitin modified vinyl ester prosthetic composite on thermal and water accelerated aging conditions. *Polymer Composites* 45 (6) (2024): 5111–5124.

Kolathayar, Sreevalsa, C. A. Aravind, and Sanjay Kumar Shukla. Performance assessment of areca leaf grids for soil reinforcement applications. Journal of Natural Fibers 18 (6) (2021): 813–822.

Kousaalya, Adhimoolam Bakthavachalam, Bopaiah Ittira Biddappa, Kelly Krumm, Sai Aditya Pradeep, and Srikanth Pilla. Poly (lactic acid)/areca fiber laminate composites processed via film stacking technique. *Journal of Applied Polymer Science* 135 (5) (2018): 45795.

Krishnamoorthy, N., N. Nagabhooshanam, Pothamsetty Kasi V. Rao, Rajesh Verma, D. Sendil Kumar, Gullapalli Ajay Sankar, Boddepalli Kiran Kumar, and V. Mohanavel. A characterization study on toughening natural fibre composites using functionalized barely husk biosilica. *Biomass Conversion and Biorefinery* (2023): 1–8.

Krishnan, G. Sai, L. S. Jayakumari, L. Ganesh Babu, and G. Suresh. Investigation on the physical, mechanical and tribological properties of areca sheath fibers for brake pad applications. Materials Research Express 6 (8) (2019): 085109.

Kumar, Sunil. Tobacco and areca nut chewing—Reproductive impairments: An overview. *Reproductive Toxicology* 36 (2013): 12-17.

Kumar, SM Sanjay, and R. Naveen Kumar. Tribological behaviour of areca fiber reinforced with hybrid composites. *Materials Today: Proceedings* 46 (2021c): 4840–4846.

Kumar, R. Nandha, and G. Parthipan. Optimization of reinforcement characteristics on eco-friendly polymer composite rebar building material developed using domestically discarded vegetable waste biochar and areca chopped fiber. *Biomass Conversion and Biorefinery* 14 (17) (2024): 21497–21505.

Kumar, GCA, Yadav D. Bodke, B. Manjunatha, and O. Nagaraja. Novel Areca Nut extract mediated preparation of TiO2 nanoparticles and its catalytic synthesis of 3, 3′-(phenyl methylene) bis-(4-hydroxy-2H-chromen-2-one) derivatives. *Chemical Data Collections* 38 (2022): 100832.

Kumaran, Keerthana, Akshara Ravikumar, Sharmila Saminathan, and Janarthanan Balasundaram. Green Synthesis of Zinc Oxide Nanoparticles Using Areca catechu. In *Encyclopedia of Green Materials*, pp. 1–8. Singapore: Springer Nature Singapore, 2023.

Lakshmana Kishore, T., and Kuma Gowwomsa Erko. Mechanical properties of epoxy composite using papaya slice biochar and areca nut chopped fibre. *Advances in Polymer Technology* 2022 (1) (2022): 4733375.

Lalhmunsiama, Lee, Seung Mok, Suk Soon Choi, and Diwakar Tiwari. Simultaneous removal of Hg (II) and phenol using functionalized activated carbon derived from areca nut waste. Metals 7 (7) (2017): 248.

Lazim, Y., L., S. M. Sapuan, Edi Syams Zainudin, and M. Mariatti. Characterization of physical, mechanical, thermal and morphological properties of agro-waste betel nut (Areca catechu) husk fibre. *Journal of Cleaner Production* 72 (2014a): 174–180.

Lazim, Yusriah, Mohd Sapuan Salit, Edi Syams Zainudin, Mariatti Mustapha, and Mohammad Jawaid. Effect of alkali treatment on the physical, mechanical, and morphological properties of waste betel nut (Areca catechu) husk fibre. BioResources 9 (4) (2014b): 7721–7736.

Lazim et al. (2014c). Nut husk alkaline treatment, BioResources 9(4), 7721–7736.

Le, Phuoc-Anh, Van-Truong Nguyen, Sumanta Kumar Sahoo, Tseung Yuen Tseng, and Kung-Hwa Wei. "Porous carbon materials derived from areca palm leaves for high performance symmetrical solid-state supercapacitors." Journal of Materials Science 55, no. 24 (2020): 10751–10764.

Lee, Kang Pa, Giftania Wardani Sudjarwo, Ji-Su Kim, Septrianto Dirgantara, Won Jai Maeng, and Heeok Hong. The anti-inflammatory effect of Indonesian Areca catechu leaf extract in vitro and in vivo. Nutrition Research and Practice 8 (3) (2014): 267–271.

Li, Xiao-ming, Wei Zheng, Dong-bo Wang, Qi Yang, Jian-bing Cao, Xiu Yue, Ting-ting Shen, and Guang-ming Zeng. Removal of Pb (II) from aqueous solutions by adsorption onto modified areca waste: Kinetic and thermodynamic studies. Desalination 258 (1-3) (2010): 148–153.

Li, Yi-Chen, Ann-Joy Cheng, Li-Yu Lee, Yu-Chen Huang, and Joseph Tung-Chieh Chang. Multifaceted mechanisms of areca nuts in oral carcinogenesis: the molecular pathology from precancerous condition to malignant transformation. *Journal of Cancer* 10 (17) (2019): 4054.

Li, Yao, Shan Yao, Yang Chen, Long Wu, Dong Xiang, and Wei-min Zhang. Synthesis and characterization of zinc ion-integrated quercetin delivery system using areca nut seeds nanocellulose. *LWT* 192 (2024): 115673.

Limaye, Akshay S., Pranita Rananaware, Arnab Ghosh, Thrilokraj Rajashekarreddy, Nandini Raghavendrarao, Varsha Brahmkhatri, Rajeev V. Hegde, and Ramesh B. Dateer. Greener Approach for Synthesis of δ-MnO_2 Nanoparticles: Access to Pharmaceutically Important Pyrimidines and their Antimicrobial Activity Studies. *ACS Applied Bio Materials* 7 (3) (2024): 1790-1800.

Lin, Cheng Jung, Salim Hiziroglu, Shu Min Kan, and Hsien Wen Lai. Manufacturing particleboard panels from betel palm (Areca catechu Linn.). *Journal of materials processing technology* 197 (1–3) (2008): 445–448.

Liu, Pei-Feng, and Yung-Fu Chang. The Controversial Roles of Areca Nut: Medicine or Toxin?. *International Journal of Molecular Sciences* 24 (10) (2023): 8996.

Liu, Huiling, Hao Zheng, Jiachao Zhang, Fang Chen, Xiaosong Hu, and Xiaofei Wang. Review of the toxic effects and health functions of arecoline on multiple organ systems. *Food Innovation and Advances* 3 (1) (2024): 31–41.

Loganathan, Tamil Moli, Mohamed Thariq Hameed Sultan, Mohammad Jawaid, Ain Umaira Md Shah, Qumrul Ahsan, Manohar Mariapan, and Mohd Shukry bin Abdul Majid. Physical, thermal and mechanical properties of areca fibre reinforced polymer composites—an overview. *Journal of Bionic Engineering* 17 (2020): 185–205.

Lokesh, Kadambari, Luana Ladu, and Louise Summerton. Bridging the gaps for a 'circular'bioeconomy: Selection criteria, bio-based value chain and stakeholder mapping. *Sustainability* 10 (6) (2018): 1695.

Lu, Hsuan-Hsuan, Shou-Yen Kao, Tsung-Yun Liu, Shou-Tien Liu, Wei-Pang Huang, Kuo-Wei Chang, and Shu-Chun Lin. Areca nut extract induced oxidative stress and upregulated hypoxia inducing factor leading to autophagy in oral cancer cells. *Autophagy* 6 (6) (2010): 725-737.

Luong, Jasper Hoi Chun, Zisis Kozlakidis, Io Hong Cheong, and Hui Wang. Innovations and limitations in areca nut research: a narrative review. *Innovations in Digital Health, Diagnostics, and Biomarkers* 3 (2023): 9–14.

Machová, Martina, Tomáš Bajer, David Šilha, Karel Ventura, and Petra Bajerová. Volatiles composition and antimicrobial activities of areca nut extracts obtained by simultaneous distillation–extraction and headspace solid-phase microextraction. *Molecules* 26 (24) (2021): 7422.

Madu, Kingsley, George O. Okoronkwo, and Emmanuel I. Nwankwo. Effects of fiber volume fraction and curing time on impact–hardness strength properties of areca fibers reinforced polyester thermoset composites. An International Open Free Access, Peer Reviewed Research Journal ISSN (2019): 0184–7937.

Madyaratri, Elvara Windra, Muhammad Rasyidur Ridho, Apri Heri Iswanto, Linda Makovická Osvaldová, Seng Hua Lee, Petar Antov, and Widya Fatriasari. Effect of Lignin or Lignosulfonate Addition on the Fire Resistance of Areca (Areca catechu) Particleboards Bonded with Ultra-Low-Emitting Urea-Formaldehyde Resin. *Fire* 6 (8) (2023): 299.

Maheswari, P. Uma, M. Saranya, K. Agnes, and M. Kanchana. Composting of Areca nut leaf sheath and its effects on growth and biochemical contents of Vignaunguiculata L. International Research Journal of Public and Environmental Health 2(2), 23–26, 2015. https://doi.org/10.15739/irjpeh.015

Mahmud, Rois Uddin, Mir Shahidul Islam, Shahin Hossain, Jannatul Nasrin, and Ayub Nabi Khan. Agro-waste areca nut husk and bagasse fiber reinforced epoxy-based hybrid composite for thermal insulated false ceiling application. *SPE Polymers* 6 (1) (2025): e10166.

Makruf, Zulhan Insan, Wanda Afnison, and Bulkia Rahim. A Study on the utilization of areca nut husk fiber as a natural fibre reinforcement in composite applications: A systematic literature review. Journal of Engineering Researcher and Lecturer 3 (1) (2024): 18–28.

Mandal, Sujata, Dayana Stephen, and Sreeram Kalarical Janardhanan. Activated carbon with composite pore structures made from peanut shell and areca nut fibers as sustainable adsorbent material for the efficient removal of active pharmaceuticals from aqueous media. *RSC Sustainability* 2 (10) (2024): 3022–3035.

Manjunath, Balasubramanya, Claudiane M. Ouellet-Plamondon, B. B. Das, Subba Rao, Chandrasekhar Bhojaraju, and Manu Rao. Areca nut husk biochar as a sustainable carbonaceous filler for cement: Pyrolysis temperature and its effect on characterization, strength, and hydration. *Industrial Crops and Products* 222 (2024): 119883.

Manjunatha, G. B., K. N. Bharath, H. M. Harsha, and P. Sivaprakash. Mode-I fracture toughness evaluation of bio based short areca fibers reinforced polymer composites. In *IOP Conference Series: Materials Science and Engineering*, 925 (1), p. 012010. IOP Publishing, 2020.

Mansingh, Bright Brailson, Govindarajan Bharathiraja, Joseph Selvi Binoj, Manikandan Natarajan, Heru Suryanto, Suchart Siengchin, and Mavinkere Rangappa Sanjay. Influence of optimal alkali treated Areca catechu L. peduncle fiber for light weight polymer composites applications. Journal of Applied Polymer Science 141 (17) (2024): e55268.

Matan, Narumol, WarasriSaengkrajang, and NirundornMatan. Antifungal activities of essential oils applied by dip-treatment on areca palm (Areca catechu) leaf sheath and persistence of their potency upon storage. International Biodeterioration& Biodegradation 65 (1), (2011): 212–216.

Matan, Narumol, MudtorlepNisoa, NirundornMatan, and TanongAewsiri. Effect of cold atmospheric plasma on antifungal activities of clove oil and eugenol against molds on areca palm (Areca catechu) leaf sheath. International Biodeterioration& Biodegradation 86 (2014): 196–201.

Mathapati, Vishal C., N. Muralidhar, B. R. Tulasiram, and Muttana S. Balreddy. Experimental study on dynamic mechanical properties of arecanut husk fiber composite. Materials Today: Proceedings 66 (2022): 858–865.

Mathew, Agnus T., Vinay S. Bhat, S. Supriya, T. Maiyalagan, Anitha Varghese, and Gurumurthy Hegde. TEMPO mediated electrocatalytic oxidation of pyridyl carbinol using palladium nanoparticles dispersed on biomass derived porous nanoparticles. *Electrochimica Acta* 354 (2020): 136624.

Mehrtash, Hedieh, Kalina Duncan, Mark Parascandola, Annette David, Ellen R. Gritz, Prakash C. Gupta, Ravi Mehrotra et al. Defining a global research and policy agenda for betel quid and areca nut. *The Lancet Oncology* 18 (12) (2017): e767–e775.

Miah, Md Jihad, Ye Li, Suvash Chandra Paul, Adewumi John Babafemi, and Jeong Gook Jang. Mechanical strength, shrinkage, and porosity of mortar reinforced with areca nut husk fibers. *Construction and Building Materials* 363 (2023): 129688.

Mishra, Ranjeet Kumar, and Kaustubha Mohanty. Fuel properties and compositional analysis of Areca catechu sawdust over MgO and ZSM-5 catalysts. Journal of the Energy Institute 94 (2021): 252–262.

Mishra, Ranjeet Kumar, Bhavana Gariya, Priyanka Savvasere, Devanshu Dhir, Pradeep Kumar, and Kaustubha Mohanty. Thermocatalytic pyrolysis of waste areca nut into renewable fuel and value-added chemicals. *ACS omega* 9 (24) (2024): 25779–25792.

Mitra, Sudip, Aman Chourasia, Ashmita Das, Nihal Gujre, and Ankit Soni. Areca nut husk biochar prepared through a low-cost kiln for the adsorptive elimination of nickel and manganese from water: a farmer-friendly approach for agro-residue management. *Biomass Conversion and Biorefinery* (2024): 1–13.

Mohanty, DebapriyaPinaki, AnirudhUdupa, Anil Chandra AR, Koushik Viswanathan, James B. Mann, Kevin P. Trumble, and Srinivasan Chandrasekar. Mechanical Behavior and High Formability of Palm Leaf Materials. *Advanced Energy and Sustainability Research* 2 (4) (2021): 2000080.

Mohanty, Debapriya Pinaki, James B. Mann, Anirudh Udupa, A. R. Anil Chandra, and Srinivasan Chandrasekar. Improving formability of palm leaf materials for foodware manufacturing using sodium hydroxide treatment. *MRS Communications* 12 (6) (2022): 1235–1243.

Mollah, Mohammad Zahirul Islam, Md Sumon Miah, Md Washim Akram, Sayed Hasan Mahmud, Mohammad Rashed Iqbal Faruque, and Kholoud Saad Al-mugren. Thermoplastic-polymer matrix composite of banana/betel nut husk fiber reinforcement: Physico-mechanical properties evaluation. *e-Polymers* 24 (1) (2024): 20230158.

Mondal, Suvankar, Sanjib Bhattacharya, and Moulisha Biswas. Antidiabetic activity of Areca catechu leaf extracts against streptozotocin induced diabetic rats. Journal of Advanced Pharmacy Education & Research 2 (1) (2012): 10-17.

Monisha, Saravanan, Rajendran K. Selvakesavan, Reena Navaroja, Chinnasamy Krishnaveni, and Marimuthu Kanchana. Utilization of arecanut leaf sheath and orange peel as environmentally friendly alternative substrates for cultivating Pleurotus ostreatus. *Biomass Conversion and Biorefinery* (2024): 1–12.

Munshi, Md Rezwan, and Md Abdul Gafur. Effect of fiber ratio on mechanical, thermal and electrical properties of areca sheath–bamboo fiber reinforced hybrid polyester composites. *Advances in Materials and Processing Technologies* 8 (4) (2022): 3975–3989.

Muralidhar, N., Vadivuchezhian Kaliveeran, V. Arumugam, and I. Srinivasula Reddy. A study on areca nut husk fibre extraction, composite panel preparation and mechanical characteristics of the composites. *Journal of The Institution of Engineers (India): Series D* 100 (2019): 135–145.

Muralidhar, N., K. Vadivuchezhian, V. Arumugam, and I. Srinivasula Reddy. Flexural modulus of epoxy composite reinforced with Arecanut husk fibre (AHF): A mechanics approach. *Materials Today: Proceedings* 27 (2020): 2265–2268.

Muralidhar, N., Kaliveeran Vadivuchezhian, V. Arumugam, and I. Srinivasula Reddy. Dynamic mechanical characterization of epoxy composite reinforced with areca nut husk fiber. *Archive of Mechanical Engineering* (2020): 57–72.

Muralidharan, Kandasamy, Ganapathi Arumugam, Amjad A. Pasha, and Nazrul Islam. Influence of areca nutshell-reduced graphene oxide, isopropanol, and exhaust gas recirculation in an internal combustion engine. *ACS omega* 7 (45) (2022): 40815-40825.

Muslim, A., S. Aprilia, T. A. Suha, and Z. Fitri. Adsorption of Pb (II) ions from aqueous solution using activated carbon prepared from areca catechu shell: Kinetic, isotherm and thermodynamic studies. Journal of the Korean Chemical Society 61 (3) (2017): 89-96.

Muthukumar, N., and G. Thilagavathi. Development of areca husk fiber/low melt PET needle punched nonwovens for building insulation applications. *Materials Today: Proceedings* (2023).

Nagaraja, R., B. R. Gurumurthy, and M. B. Shivanna. Bio softening of arecanut waste areca husk, leaf and leaf sheath for value added compost. Int. J. Res. Appl. Nat. Soc. Sci 2 (2014): 105–112.

Najm, Asmaa Soheil, Norasikin A. Ludin, Mahir Faris Abdullah, Munirah A. Almessiere, Naser M. Ahmed, and Mahmoud AM Al-Alwani. Areca catechu extracted natural new sensitizer for dye-sensitized solar cell: Performance evaluation. *Journal of Materials Science: Materials in Electronics* 31 (2020): 3564–3575.

Narayanamurthy, G., Y. L. Ramachandra, S. Padmalatha Rai, Y. N. Manohara, and B. T. Kavitha. Areca husk: An inexpensive substrate for citric acid production by Aspergillus niger under solid state fermentation. (2008).

Naveenkumar, K. J., B. Thippeswamy, B. V. Thirumalesh, and S. P. Banakar. Effect of pretreatment for delignification and sugar recovery for bioethanol production from Areca nut husk waste. International Journal of Current Research 4 (1) (2012): 18-24.

Nayak, Subhakanta, and JyotiranjanMohanty. Erosion wear behavior of benzoyl chloride modified areca sheath fiber reinforced polymer composites. *Composites Communications* 18 (2020): 19–25.

Nayak, Subhakanta, and Jyoti Ranjan Mohanty. Influence of chemical treatment on tensile strength, water absorption, surface morphology, and thermal analysis of areca sheath fibers. *Journal of Natural Fibers* 16 (4) (2019): 589–599.

Nayak, Subhakanta, and Jyoti Ranjan Mohanty. Study of mechanical, thermal, and rheological properties of areca fiber-reinforced polyvinyl alcohol composite. Journal of Natural Fibers 16 (5) (2019): 688–701.

Nayak, Subhakanta, and Jyotiranjan Mohanty. Two-body abrasive behavior of areca sheath fibre reinforced polyvinyl alcohol composites. Indian Journal of Fibre & Textile Research (IJFTR) 44 (2) (2019): 205–209.

Nayak, Subhakanta, Jyoti Ranjan Mohanty, Priya Ranjan Samal, and Bijoy Kumar Nanda. Polyvinyl chloride reinforced with areca sheath fiber composites—An experimental study. Journal of Natural Fibers 17 (6) (2020): 781–792.

Nayak, S., and S. D. Mohanty. Solid Particle Erosion Properties of Areca Sheath Fiber Reinforced Polyvinyl Chloride Composites. https://www.researchgate.net/profile/Saumya-Mohanty/publication/344041215_Solid_Particle_Erosion_Properties_of_Areca_Sheath_Fiber_Reinforced_Polyvinyl_Chloride_Composites/links/5f4f397792851c250b88c8d6/Solid-Particle-Erosion-Properties-of-Areca-Sheath-Fiber-Reinforced-Polyvinyl-Chloride-Composites.pdf

Nayak, Subhakanta, Sujit Kumar Khuntia, SaumyaDarsanMohanty, JagannathMohapatra, and Tapan Kumar Mall. An Experimental Study of Physical, Mechanical and Morphological Properties of Alkali Treated Moringa/areca Based Natural Fiber Hybrid Composites. Journal of Natural Fibers (2020): 1–12.

Nayak, Subhakanta, Sujit Kumar Khuntia, Saumya Darsan Mohanty, Jagannath Mohapatra, and Tapan Kumar Mall. An experimental study of physical, mechanical and morphological properties of alkali treated moringa/areca based natural fiber hybrid composites. Journal of Natural Fibers 19 (2) (2022): 630–641.

Nayak, Subhakanta, and Jyoti R. Mohanty. Optimization of operating parameters for abrasion of areca sheath reinforced polymer composite: grey based Taguchi approach. International Journal of Plastics Technology 22 (1) (2018): 26-40.

Neelappa, Deviprasad Renjadi, Suryanarayana Keerikadu, and Lokesh K. Sri Ramamurthy. Fabrication and characterization of bio composite fiber boards from areca leaf sheaths. *Journal of Materials and Manufacturing* 2 (1) (2023): 44–53.

Neelappa, Deviprasad Renjadi, Suryanarayana Keerikadu, and K. S. Lokesh. Acoustic and Thermal Analysis of Composite Fibre Boards Fabricated Using Areca Catechu & Areca Husk Fibers. *Journal of The Institution of Engineers (India): Series D* (2024): 1–12.

Nguyen, Tien Dung, Thuy Chinh Nguyen, Xuan Thai Nguyen, Thi Kim Anh Nguyen, Thao Linh Bui, Thanh Thuy Tran, and Hoang Thai. Eco-friendly synthesis of hydrotalcite-Ag nanosheets using Areca catechu L. nut extract and its antibacterial activity. Journal of Science: Advanced Materials and Devices 9 (2) (2024): 100678.

Nonthakaew, Athiya, NarumolMatan, TanongAewsiri, and NirundornMatan. Antifungal activity of crude extracts of coffee and spent coffee ground on areca palm leaf sheath (Areca catechu) based food packaging. Packaging technology and science 28 (7) (2015): 633–645.

Novera, Tamanna Mamun, Mumtahina Tabassum, Mondira Bardhan, Md Azharul Islam, and Md Atikul Islam. Chemical modification of betel nut husk prepared by sodium hydroxide for methylene blue adsorption. Applied Water Science 11 (4) (2021): 1–14.

Nurdin, Hendri, and Andre Kurniawan. Particle board made from areca fiber with tapioca adhesive. In *Journal of Physics: Conference Series*, 1594 (1), p. 012031. IOP Publishing, 2020.

Padmaraj, N. H., Laxmikant G. Keni, K. N. Chetan, and Mayur Shetty. "Mechanical characterization of areca husk-coir fiber reinforced hybrid composites." *Materials Today: Proceedings* 5 (1) (2018): 1292–1297.

Pai, Apoorva M., Mahesh M. Shanbhag, Thandavarayan Maiyalagan, Sondos Abdullah Alqarni, and Nagaraj P. Shetti. Activated carbon synthesized from Arecanut catechu L. as a sustainable precursor intercalated TiO_2 modified electrode for the detection of fungicide Dichlorophen. *Diamond and Related Materials* 140 (2023): 110561.

Palanichamy, Kalyani, and Anitha Ariharaputhiran. Areca leaves as a source of carbon: Preliminary investigation as catalyst support for electrolytic hydrogen evolution in acidic medium. International journal of hydrogen energy 38 (5) (2013): 2263-2270.

Palanisamy, Sivasubramanian, S. Keerthiveetil Ramakrishnan, Carlo Santulli, Tabrej Khan, and Omar Shabbir Ahmed. Mechanical and wear performance evaluation of natural fiber/epoxy matrix composites. *BioResources* 19 (4) (2024): 845.

Pant, Bishnu Datt, Deepa Neupane, Dasu Ram Paudel, Prakash Chandra Lohani, Surendra Kumar Gautam, Megh Raj Pokhrel, and Bhoj Raj Poudel. Efficient biosorption of hexavalent chromium from water by modified arecanut leaf sheath. *Heliyon* 8 (4) (2022).

Paolillo, Vincenzo, Mahendran Jayakumar, Colton Sheperd, Andrew Tran, Stephanie Hoang, Nhu Dao, Parag Jain, and Alan L. Myers. In vitro hydrolysis of areca nut xenobiotics in human liver. *Drug metabolism and pharmacokinetics* 60 (2025): 101039.

Park, Yong Bok, Seon-Min Jeon, Sung-June Byun, Hee-Sook Kim, and Myung-Sook Choi. Absorption of intestinal free cholesterol is lowered by supplementation of Areca catechu L. extract in rats. *Life sciences* 70 (16) (2002): 1849–1859.

Pathania, Dimple, Ankita Araballi, Fiona Fernandes, Jyothi Mannekote Shivanna, Ganesan Sriram, Mahaveer Kurkuri, Gurumurthy Hegde, and Tejraj M. Aminabhavi. Cost effective porous areca nut carbon nanospheres for adsorptive removal of dyes and their binary mixtures. *Environmental Research* 224 (2023): 11552.

Paul, R. Christu, Balaji Ramachandran, Gullipalli Sushma, KH Aditya Harshavardhan, and Isukapalli Rohith. An empirical research on areca fiber polymer composite for automotive components in modern industry. *Materials Today: Proceedings* 33 (2020): 4493–4497.

Paul, Shantanu, Mohammad Shariful Islam, and Namrata Chakma. Effectiveness of areca fiber and cement on the engineering characteristics of compressed stabilized earth blocks. *Construction and Building Materials* 427 (2024): 136290.

Peng, Wei, Yu-Jie Liu, Na Wu, Tao Sun, Xiao-Yan He, Yong-Xiang Gao, and Chun-Jie Wu. Areca catechu L.(Arecaceae): A review of its traditional uses, botany, phytochemistry, pharmacology and toxicology. *Journal of ethnopharmacology* 164 (2015): 340–356.

Perumal, Anand Babu, Reshma B. Nambiar, Periyar Selvam Sellamuthu, Emmanuel Rotimi Sadiku, Xiaoli Li, and Yong He. Extraction of cellulose nanocrystals from areca waste and its application in eco-friendly biocomposite film. Chemosphere 287 (2022): 132084.

Poddar, P., S. Sultana, and M. A. Akbar. Environmentally sustainability of short Areca-nut leaf sheath fiber reinforced polypropylene composites. MOJ Poly Sci 2 (2) (2018): 48-50.

Poddar, Pinku, Mohammad Asadulah Asad, Muhammad Saiful Islam, Shahin Sultana, Husna Parvin Nur, and AM Sarwaruddin Chowdhury. Mechanical and morphological study of arecanut leaf sheath (ALS), coconut leaf sheath (CLS) and coconut stem fiber (CSF). Advanced Materials Science 1 (2) (2016): 1–4.

Poulose, Aiswarya, Ajith Mathew, Arya Uthaman, Hiran Mayookh Lal, Jyotishkumar Parameswaranpillai, A. Mathiazhagan, Mohamed Mohamed Saheed et al. Facile fabrication of arecanut palm sheath based robust hydrophobic cellulose nanopapers via self-assembly of ZnO nanoflakes and its shelf-life prediction for sustainable packaging applications. International Journal of Biological Macromolecules 255 (2024): 128004.

Prabakaran, J., S. Santhosh, C. Saravanakumar, R. Selva Rabins, and S. Sudhanraj. Investigation of mechanical properties in Areca (betel nut) and sisal fiber with epoxy composite. *Journal for Researchl Volume* 3 (2) (2017).

Prabhu, Rachana V., Vishnudas Prabhu, Laxmikanth Chatra, Prashant Shenai, Nithin Suvarna, and Savita Dandekeri. Areca nut and its role in oral submucous fibrosis. *Journal of clinical and experimental dentistry* 6 (5) (2014): e569.

Pradeep, Bhavana, Priyanka Hemba, Anusha Kunnathuli Jagadeesh, Chaitra Ganappa Ramakkanavar, Sneha Nayak, and Chandrayan Vaman Rao. Anticandidal and antioxidant activity of silver and gold nanoparticles biosynthesized using matured areca nut husk extract. *International Journal of Nanoparticles* 13 (1) (2021): 21–32.

Prakash, T., Sumanth Ratna Kandavalli, Santosh Kumar Sahu, and T. Mayavan. Development of flexible poly vinyl alcohol composite for electromagnetic interference shielding using areca microfiber, cobalt Nps, and rice bran biochar. *Biomass Conversion and Biorefinery* (2024): 1–11.

Prasannakumaran, K. M., Mohammed Abdullah Bareen, Jatindra K. Sahu, and Vijayaraghavan M. Chariar. Areca sheath industrial waste valorization: resource mapping, physiochemical parameters, and heavy metal analysis. *Biomass Conversion and Biorefinery* (2024): 1–15.

Praveena, B. A., N. Santhosh, Abdulrajak Buradi, H. V. Srikanth, G. Shankar, K. Ramesha, N. Manjunath, S. N. Karthik, M. Rudra Naik, and S. Praveen Kumar. Experimental investigation on density and volume fraction of void, and mechanical characteristics of areca nut leaf sheath fiber-reinforced polymer composites. International Journal of Polymer Science 1 (2022): 6445022.

Raghavendra, Narasimha, and Jathi Ishwara Bhat. Anti-corrosion properties of areca palm leaf extract on aluminium in 0.5 M HCl environment. South African Journal of Chemistry 71 (2018): 30–38.

Raghavendra, Vinay B., Sushmitha Shankar, M. Govindappa, Arivalagan Pugazhendhi, Minaxi Sharma, and Siddaiah Chandra Nayaka. Green synthesis of zinc oxide nanoparticles (ZnO NPs) for effective degradation of dye, polyethylene and antibacterial performance in waste water treatment. *Journal of Inorganic and Organometallic Polymers and Materials* (2022): 1–17.

Raghupathy, R., R. Viswanathan, and C. T. Devadas. Quality of paper boards from arecanut leaf sheath. Bioresource technology 82 (1) (2002): 99-100.

Rahman, Ave Olivia, and Alfi Khatib. Formulation and characterization of 50% ethanol extract of areca nut (Areca catechu) nanoparticles using the ionic gelation method. Medical Sains: Jurnal Ilmiah Kefarmasian 8 (3) (2023): 981–986.

Rahman, M. Arifur, Papeya Sultana, M. Sahidul Islam, M. Toslim Mahmud, M. Mamun Or Rashid, and Foysal Hossen. Comparative antimicrobial activity of Areca catechu nut extracts using different extracting solvents. *Bangladesh Journal of Microbiology* (2014): 19–23.

Rahman, Fazlar, Ismail Hossain Rafi, Sadib Chowdhury, Safaul Islam Rohan, Md Wahid-Saruar, Mohammad Rejaul Haque, M. A. Gafur, and Mahbub Hassan. Effect of fiber orientation on mechanical properties of betel nut (areca palm) stem fiber reinforced laminated polyester composites. *Advances in Materials and Processing Technologies* (2024): 1–18.

Rahman, Fazlar, Ismail Hossain Rafi, Sadib Chowdhury, Safaul Islam Rohan, Md Wahid-Saruar, Mohammad Rejaul Haque, M. A. Gafur, and Mahbub Hassan. Effect of fiber orientation on mechanical properties of betel nut (areca palm) stem fiber reinforced laminated polyester composites. *Advances in Materials and Processing Technologies* 11 (2) (2025): 844–861.

Raj, Manish, Shahab Fatima, and Naresh Tandon. A study of areca nut leaf sheath fibers as a green sound-absorbing material. Applied Acoustics 169 (2020): 107490.

Raj, Sonia S., J. Edwin Raja Dhas, and B. Harish Kumar. Optimization and Analysis of Abrasive Wear of Agro-waste Fiber Reinforced Composites by RSM Design Matrix. In *Recent Advances in*

Materials Technologies: Select Proceedings of ICEMT 2021, pp. 111–130. Singapore: Springer Nature Singapore, 2022a.

Raj, Sonia S., J. Edwin Raja Dhas, and B. Harish Kumar. Tribological Applications of Agro Waste: Areca Husk Fibre Polymer Composites. (2022b).

Rajan, Anish, Vidya Vilas, and Daizy Philip. Studies on catalytic, antioxidant, antibacterial and anticancer activities of biogenic gold nanoparticles. *Journal of Molecular Liquids* 212 (2015): 331-339.

Rajan, Anish, Vidya Vilas, and Daizy Philip. Catalytic and antioxidant properties of biogenic silver nanoparticles synthesized using Areca catechu nut. *Journal of Molecular Liquids* 207 (2015): 231-236.

Rajan, Akhila, Jayalakshmi Gopinatha Kurup, and Tholath Emilia Abraham. Solid state production of manganese peroxidases using arecanut husk as substrate. Brazilian Archives of Biology and Technology 53 (3) (2010): 555–562.

Rajan, Akhila, Jayalakshmi Gopinadha Kurup, and Tholath Emilia Abraham. Biosoftening of arecanut fiber for value added products. Biochemical Engineering Journal 25 (3) (2005): 237–242.

Rajbongshi, Bedanta, Akuleti Saikumar, and Laxmikant Shivnath Badwaik. Valorization of areca nut husk and water hyacinth fibers into biodegradable plates for sustainable packaging. *Sustainable Food Technology* (2025).

Rajesh, Murkur, and K. Balaji. "Evaluation properties of areca fiber reinforced biodegradable composite material." *Materials Today: Proceedings* (2023). Evaluation properties of areca fiber reinforced biodegradable composite material – ScienceDirect.

Rajeshkannan, S., Annamalai Sivakumar, M. Mariappan, and J. Gunasekaran. Performance Evaluation of PET Foam and Areca Fruit Fibre Reinforced Modified Cellulose-Vinyl Ester Composite on Thermal and Water Ageing Conditions. *Waste and Biomass Valorization* (2024): 1–15.

Ramasubbu, Rajkumar, Arumugam Kayambu, Sivasubramanian Palanisamy, and Nadir Ayrilmis. Mechanical Properties of Epoxy Composites Reinforced with Areca catechu Fibers Containing Silicon Carbide. *BioResources* 19 (2) (2024).

Ramesh, Shruthy, and Preetha Radhakrishnan. Areca nut fiber nano crystals, clay nano particles and PVA blended bionanocomposite material for active packaging of food. *Applied Nanoscience* 12 (3) (2022): 295-307.

Rangaraj, R., S. Sathish, T. L. D. Mansadevi, R. Supriya, Raviteja Surakasi, M. Aravindh, Alagar Karthick et al. Investigation of Weight Fraction and Alkaline Treatment on Catechu Linnaeus/Hibiscus cannabinus/Sansevieria Ehrenbergii Plant Fibers-Reinforced Epoxy Hybrid Composites. *Advances in Materials Science and Engineering* 1 (2022): 4940531.

Rashid, Md Hasibur, Mahabubur Rahman Sakil, Nakib Rayhan, and Hriday Ahmed. Study on the Effects of Incorporating Areca Fibre on mechanical Properties of Natural Fibre Reinforced Concrete. Dinkum Journal of Natural & Scientific Innovations, 3 (2) (2024).

Reddy, Narendra, and Yiqi Yang. Preparation and characterization of long natural cellulose fibers from wheat straw. *Journal of Agricultural and Food Chemistry* 55 (2)1 (2007): 8570–8575.

Reddy, Narendra, and Yiqi Yang. Natural cellulose fibers from soybean straw.*Bioresource Technology* 100 (14) (2009): 3593–3598.

Reddy, Narendra, and Yiqi Yang. Structure and properties of high quality natural cellulose fibers from cornstalks. *Polymer* 46 (15) (2005): 5494-5500.

Reddy, Narendra, and Yiqi Yang. Properties and potential applications of natural cellulose fibers from cornhusks. *Green Chemistry* 7 (4) (2005): 190-195.

Reddy, Narendra, and Yiqi Yang. Natural cellulose fibers from switchgrass with tensile properties similar to cotton and linen. *Biotechnology and bioengineering* 97 (5) (2007): 1021-1027.

Renugadevi, N., Anitha G., Lalitha P. Kinetics of the removal of hexavalent chromium using a low cost activated carbon adsorbent, Advances in Applied Science Research, 1(2), 2010, 102-105.

Ridho, Muhammad Rasyidur, Deded Sarip Nawawi, Ika Juliana, and Widya Fatriasari. The kraft lignin characteristics of areca leaf sheath isolated by phosphoric acid. *Bioresource Technology Reports* 23 (2023): 101569.

Ridho, Muhammad Rasyidur, Muhammad Adly Rahandi Lubis, Deded Sarip Nawawi, and Widya Fatriasari. Optimization of areca leaf sheath nanolignin synthesis by a mechanical method for in situ modification of ultra-low molar ratio urea-formaldehyde adhesives. *International Journal of Biological Macromolecules* 271 (2024b): 132614.

Rumgay, Harriet, Suzanne T. Nethan, Richa Shah, Jérôme Vignat, Olalekan Ayo-Yusuf, Pankaj Chaturvedi, Eliete NS Guerra et al. Global burden of oral cancer in 2022 attributable to smokeless tobacco and areca nut consumption: a population attributable fraction analysis. *The Lancet Oncology* 25 (11) (2024): 1413–1423.

Saha, Dipankar, Bidesh Roy, and Patit Paban Kundu. Areca nut husk nanoadditive for compression ignition engine: characterisation, energy–exergy–exergoeconomic, and sustainability analyses. *Journal of Thermal Analysis and Calorimetry* 149 (21) (2024): 12385–12398.

Saha, Dipankar, and Bidesh Roy. Influence of areca nut husk nano-additive on combustion, performance, and emission characteristics of compression ignition engine fuelled with plastic-grocery-bag derived oil-water-diesel emulsion. Energy 268 (2023): 126682.

Salehi, Bahare, Dmitry A. Konovalov, Pascaline Fru, Petrina Kapewangolo, Gregorio Peron, Mileski S. Ksenija, Susana M. Cardoso et al. Areca catechu—From farm to food and biomedical applications. *Phytotherapy Research* 34 (9) (2020): 2140–2158.

Sampathkumar, Dhanalakshmi, Ramadevi Punyamurthy, Basavaraju Bennehalli, and Srinivasa Chikkol Venkateshappa. Effect of esterification on moisture absorption of single areca fiber. International Journal of Agriculture Sciences 4 (4) (2012): 227.

Sankarathil, Aju Jo, Raja Rosari, Vijay Santhiyagu Joseph, Sabitha Jannet, and Aswathy Ann Mathew. Chopped Areca Nut Fibers as Filler in Epoxy Matrix: Mechanical and Tribological Studies. *Trends in Sciences* 20 (12) (2023): 7155–7155.

Sankarathil, Aju Jo, R. Raja, S. J. Vijay, Sabitha Jannet, and Rittin Abraham Kurien. Structural, Thermal and Elemental Investigations in Epoxy Polymer Composites Reinforced with NaOH-Treated Short Areca Nut Fibers. *Transactions of the Indian Institute of Metals* 76 (9) (2023): 2525–2533.

Santosh, MB, Mohan Kumar GC, and Jeyaraj Pitchaimani. Acoustic characterization of natural areca catechu fiber-reinforced flexible polyurethane foam composites. Journal of Applied Polymer Science 141 (4) (2024): e54866.

Sapaat, Nurul Akmalina, Asiah Abdullah, and Sheikh Ahmad Izaddin Sheikh Mohd Ghazali. Qualitative and quantitative phytochemical analysis and antioxidant activities of Areca catechu nut husk. Journal of Academia (2022): 71–80.

Saravanan, A., P. Thirumurugan, S. Rajeshkannan, and S. Sridhar. Effect of corn cob carbon quantum dots and areca husk microfiber on EMI shielding effectiveness of flexible PVA thin film at 8–20GHz frequency bands. *Biomass Conversion and Biorefinery* 14 (16) (2024): 18827-18834.

Sari, Liza Meutia, Gus Permana Subita, and Elza Ibrahim Auerkari. Areca nut extract demonstrated apoptosis-inducing mechanism by increased caspase-3 activities on oral squamous cell carcinoma. *F1000Research* 7 (2019): 723.

Sari, Liza Meutia, Rachmi Fanani Hakim, Zaki Mubarak, and Andriyanto Andriyanto. "Analysis of phenolic compounds and immunomodulatory activity of areca nut extract from Aceh, Indonesia, against Staphylococcus aureus infection in Sprague-Dawley rats. *Veterinary World* 13 (1) (2020): 134.

Sasi, Soorya, C. Ardra Krishna, Sunish K. Sugunan, Akash Chandran, P. Radhakrishnan Nair, K. R. V. Subramanian, and Suresh Mathew. Low cost, high efficiency flexible supercapacitor electrodes made from areca nut husk nanocellulose and silver nanoparticle embedded polyaniline. *RSC advances* 11 (47) (2021a): 29564–29575.

Sathish, S., S. Aneesh, S. Naveen Vigneshwaran, EK Nithish Kumar, M. Ravichandran, and Muhammed Anaz Khan. Role of areca fibres on the mechanical behavior of natural fibre reinforced polymer composite. *Materials Today: Proceedings* (2023).

Sathyaseelan, P., K. Raja, Murali Arun Prasad, and J. Manivannan. Experimental investigation on the ballistic impact properties of areca/kenaf fiber-reinforced hybrid epoxy composite plates. *Materials Today: Proceedings* 68 (2022): 2195–2198.

Satyanarayana, Katakam, Ganesh K. Jadhav, D. R. Srinivasan, Manepalli Sailaja, A. Sunny Kumar, Pathem Uma Chaithanya, R. G. Padmanabhan, and A. Joseph Arockiam. Effects of layering variation on mechanical, thermal, and morphological properties of areca natural fiber mat reinforced epoxy biocomposites. *Biomass Conversion and Biorefinery* (2024): 1–11.

Seki, Yasemin, Figen Selli, Ümit Halis Erdoğan, Metehan Atagür, and M. Özgür Seydibeyoğlu. A review on alternative raw materials for sustainable production: novel plant fibers. *Cellulose* 29 (9) (2022): 4877–4918.

Senevirathna, Kalpani, Roshan Pradeep, Yovanthi Anurangi Jayasinghe, Shalindu Malshan Jayawickrama, Rasika Illeperuma, Saman Warnakulasuriya, and Ruwan Duminda Jayasinghe. Carcinogenic effects of areca nut and its metabolites: a review of the experimental evidence. *Clinics and practice* 13 (2) (2023): 326–346.

Setyawan, Hendrix Yulis, Budi Waluyo, and Addion Nizori. Energy potential from Areca Palm through Direct Combustion and Pyrolysis in Indonesia: A review. Indonesian Food Science & Technology Journal 4 (1) (2020): 19–26.

Shafique, Kashif, Saira Saeed Mirza, Priya Vart, Abdul Rauf Memon, Moin Islam Arain, Muhammad Farooq Tareen, and Zia Ul Haq. Areca nut chewing and systemic inflammation: evidence of a common pathway for systemic diseases. *Journal of Inflammation* 9 (2012): 1–8.

Shafique, Kashif, Mubashir Zafar, Zeeshan Ahmed, Naveed Ali Khan, Muhammad Akbar Mughal, and Fauzia Imtiaz. Areca nut chewing and metabolic syndrome: evidence of a harmful relationship. *Nutrition Journal* 12 (2013): 1–6.

Shaikh, Zeeshan Akil, and Anil S. Maheshwari. Optimization of composition parameter of Banana-Areca fiber epoxy based composite material for impact strength using Taguchi method. *Materials Today: Proceedings* (2024).

Shanmugasundaram, N., I. Rajendran, and T. J. C. P. Ramkumar. Characterization of untreated and alkali treated new cellulosic fiber from an Areca palm leaf stalk as potential reinforcement in polymer composites. *Carbohydrate Polymers* 195 (2018): 566-575.

Sharma, Mohit, Sachin C. Sarode, Gargi Sarode, and Raghu Radhakrishnan. Areca nut-induced oral fibrosis–Reassessing the biology of oral submucous fibrosis. *Journal of Oral Biosciences* 66 (2) (2024): 320-328.

Shashidhar, R., A. Swathi, K. Vasista, and S. V. Reshma. Optimization of cellulase yield from areca husk and areca sheath using Pseudomonas fluorescens. In 2013 IEEE Global Humanitarian Technology Conference: South Asia Satellite (GHTC-SAS), pp. 21–26. IEEE, 2013.

Shih, Yu-Tzu, Po See Chen, Chi-Han Wu, Yu-Ting Tseng, Yang-Chang Wu, and Yi-Ching Lo. Arecoline, a major alkaloid of the areca nut, causes neurotoxicity through enhancement of oxidative stress and suppression of the antioxidant protective system. *Free Radical Biology and Medicine* 49, no. 10 (2010): 1471–1479.

Shunmugasundaram, M., A. Praveen Kumar, N. K. Amudhavalli, and S. Sivasankar. Investigations on the Tensile and Flexural Properties of Vacuum-Infused Areca Polymer Nanocomposites. In *Advances in Lightweight Materials and Structures: Select Proceedings of ICALMS 2020*, pp. 243–251. Singapore: Springer Singapore, 2020.

Shwetha, U. R., M. S. Latha, C. R. Rajith Kumar, M. S. Kiran, and Virupaxappa S. Betageri. Facile synthesis of zinc oxide nanoparticles using novel Areca catechu leaves extract and their in vitro antidiabetic and anticancer studies. *Journal of Inorganic and Organometallic Polymers and Materials* 30 (12) (2020): 4876–4883.

Shwetha UR, Rajith Kumar CR, Kiran MS, Virupaxappa S. Betageri, Latha MS, Ravindra Veerapur, Ghada Lamraoui et al. Biogenic synthesis of NiO nanoparticles using areca catechu leaf extract and their antidiabetic and cytotoxic effects. *Molecules* 26 (9) (2021): 2448.

Siddika, Salma, and Ahmed Sharif. Processing and characterization of areca and waste nylon fiber reinforced hybrid polypropylene composites. *International Journal of Innovation and Scientific Research* 19 (2015): 319-330.

Siddika, Salma, Ahmed Sharif, and Mahbub Hasan. Effect of areca and waste nylon fiber hybridization on the properties of recycled polypropylene composites. *Journal of Natural Fibers* 19 (13) (2022): 6625-6637.

Soman, Sumesh, Asha Susan Chacko, and Vadakkethonippurathu Sivankuttynair Prasad. Semi-interpenetrating network composites of poly (lactic acid) with cis-9-octadecenylamine modified cellulose-nanofibers from Areca catechu husk. *Composites Science and Technology* 141 (2017): 65–73.

Soman, Sumesh, Asha Susan Chacko, V. S. Prasad, P. Anju, B. S. Surya, and K. Vandana. Self-assembly of oleylamine modified nano-fibrillated cellulose from areca husk fibers into giant vesicles. *Carbohydrate polymers* 182 (2018): 69–74.

Somashekhara, J., B. T. Ramesh, Vinay Belagavi, and H. Madhu. "Investigation and study of mechanical properties of areca shell fiber and palm powder natural composites." *IOSR Journal of Mechanical and Civil Engineering* 15 (6) (2018): 62–73.

Sreena, C. P. Arecanut husk: A potential feedstock for enhanced production of endoglucanase by Bacillus subtilis MU S1. Environmental Technology & Innovation 19 (2020): 100989.

R. B., C. V. Srinivasa, and B. Basavaraju. Study on morphology and mechanical behavior of areca leaf sheath reinforced epoxy composites. Advanced Composites and Hybrid Materials 3 (3) (2020): 365–374.

Srinivasa, C. V., and K. N. Bharath. Impact and hardness properties of areca fiber-epoxy reinforced composites. *J. Mater. Environ. Sci* 2 (4) (2011b): 351-356.

Srinivasa, C. V., and K. N. Bharath. Effect of alkali treatment on impact behavior of areca fibers reinforced polymer composites. *International Journal of Materials and Metallurgical Engineering* 7 (4) (2013): 240-244.

Srinivasa, C. V., A. Arifulla, N. Goutham, T. Santhosh, H. J. Jaeethendra, R. B. Ravikumar, S. G. Anil, DG Santhosh Kumar, and J. Ashish. Static bending and impact behaviour of areca fibers composites. Materials & Design 32 (4) (2011): 2469–2475.

Srinivasan, Harinei, Hariharan Arumugam, Balaji Krishnasamy, Abdul Aleem MI, Arun Murugesan, and Alagar Muthukaruppan. Desert cotton and areca nut husk fibre reinforced hybridized bio-benzoxazine/epoxy bio-composites: Thermal, electrical and acoustic insulation applications. Construction and Building Materials 363 (2023): 129870.

Subbayamma, Gujjarlapudi, and DV Rama Koti Reddy. Development of flexible PVA EMI shielding composite using areca microfiber, cobalt, and potato peel biochar particle. *Biomass Conversion and Biorefinery* 14 (16) (2024): 19815–19824.

Subramani, B. Sheeka, S. Shrihari, B. Manu, and K. S. Babunarayan. Evaluation of pyrolyzed areca husk as a potential adsorbent for the removal of Fe^{2+} ions from aqueous solutions. Journal of environmental management 246 (2019): 345–354.

Subramanyam, Suresh Poyil, Dilip Kumar Kotikula, Basavaraju Bennehalli, Atul Babbar, Sagr Alamri, Alaauldeen A. Duhduh, Ali A. Rajhi, Raman Kumar, and Ketan Kotecha. Plain-woven areca sheath fiber-reinforced epoxy composites: the influence of the fiber fraction on physical and mechanical features and responses of the tribo system and machine learning modeling. *ACS omega* 9 (7) (2024): 8019–8036.

Subran, Nikitha, Karnapa Ajit, Haribabu Krishnan, Senthilkumar Pachiyappan, and Palani Ramaswamy. Synthesis and performance of a cathode catalyst derived from areca nut husk in microbial fuel cell. Chemosphere 312 (2023): 137303.

Sukirtha, Raman, Muthukalingan Krishnan, Rajamanickam Ramachandran, Soundararajan Kamalakkannan, Palanivel Kokilavani, Devaraj SankarGanesh, Soundarapandian Kannan, and Shanmugam Achiraman. Areca catechu Linn.–Derived Silver Nanoparticles: A Novel Antitumor Agent against Dalton's Ascites Lymphoma. *International Journal of Green Nanotechnology* 3 (1) (2011): 1–12.

Sultana, Tanvir, Shahin Sultana, Husna Parvin Nur, and Md Wahab Khan. Studies on mechanical, thermal and morphological properties of betel nut husk nano cellulose reinforced biodegradable polymer composites. *Journal of Composites Science* 4 (3) (2020): 83.

Sun, Yuanyuan, Jian Feng, Wencheng Hou, Huasha Qi, and Yangyang Liu. "Comprehensive insights into areca nut: active components and omics technologies for bioactivity evaluation and quality control. *Frontiers in Pharmacology* 15 (2024b): 1407212.

Sun, Ying, Jinfeng Lu, Jiaqi Li, Peng Li, Meihui Zhao, and Guanghua Xia. Optimization of ultrasonic-assisted extraction of polyphenol from Areca nut (Areca catechu L.) seeds using response surface methodology and its effects on osteogenic activity. *Ultrasonics Sonochemistry* 98 (2023): 106511.

Sun, Huihui, Wenzhen Yu, Hu Li, Xiaosong Hu, and Xiaofei Wang. Bioactive Components of Areca Nut: An Overview of Their Positive Impacts Targeting Different Organs. *Nutrients* 16 (5), (2024): 695.

Sunny, Georgy, and T. Palani Rajan. Review on areca nut fiber and its implementation in sustainable products development. Journal of Natural Fibers 19 (12) (2022): 4747–4760.

Supriya, S., Guddekoppa S. Ananthnag, Vijayendra S. Shetti, B. M. Nagaraja, and Gurumurthy Hegde. Cost-effective bio-derived mesoporous carbon nanoparticles-supported palladium catalyst for nitroarene reduction and Suzuki–Miyaura coupling by microwave approach. *Applied Organometallic Chemistry* 34 (3) (2020): e5384.

Supriya, S., Gurwinder Singh, Rohan Bahadur, Ajayan Vinu, and Gurumurthy Hegde. Porous carbons derived from Arecanut seeds by direct pyrolysis for efficient CO 2 capture. *Emergent Materials* (2022): 1–9.

Suresh, P. S., K. Dilip Kumar, S. Dhanalakshmi, C. V. Srinivasa, and B. Basavaraju. Effect of fiber fraction on the physical and mechanical properties of short areca sheath fiber reinforced polymer composite. *Materials Today: Proceedings* 44 (2021a): 4972–4975.

Suresh, P. S., K. Dilip Kumar, S. Dhanalakshmi, and B. Basavaraju. Physical, chemical and surface morphological characterization of single areca sheath fiber. In IOP Conference Series: Materials Science and Engineering, 1065 (1), p. 012020. IOP Publishing, 2021b.

Suresh, P. S., K. Dilip Kumar, S. N. Gautham, S. Preetham, C. V. Srinivasa, and B. Basavaraju. Tribological properties of areca sheath fiber composites. *Materials Today: Proceedings* 46 (2021c): 7955–7961.

Surianarayanan, P., N. Balaji, and K. Balasubramanian. Effect of silane-treated chitosan carbohydrate polymer and tanned leather/areca fiber hybrid epoxy composites on mechanical, drop load, and fatigue properties. *Biomass Conversion and Biorefinery* 14 (16) (2024): 19093-19106.

Swamy, R. P., GC Mohan Kumar, Y. Vrushabhendrappa, and Vince Joseph. Study of areca-reinforced phenol formaldehyde composites. *Journal of reinforced plastics and composites* 23 (13) (2004): 1373–1382.

Swathy, O., P. Deepthi, Kizhakkekilikoodayil Vijayan Baiju, P. Praveen, and Anjali Paravannoor. Sustainable production of in-situ activated carbon from arecanut husk and their supercapacitor applications. *Sustainable Chemistry for the Environment* 7 (2024): 100145.

Tabassum, Mumtahina, Mondira Bardhan, Tamanna Mamun Novera, Md Atikul Islam, Ali Hadi Jawad, and Md Azharul Islam. NaOH-Activated Betel Nut Husk Hydrochar for Efficient Adsorption of Methylene Blue Dye. Water, Air, & Soil Pollution 231 (8) (2020): 1–13.

Taer, E., R. Handayani, Apriwandi, R.Taslim, Awitdrus, A. Amri, Agustino and I. Iwantono The Synthesis of Bridging Carbon Particles with Carbon Nanotubes from Areca catechu Husk Waste as Supercapacitor Electrodes. Int. J. Electrochem. Sci., 14 (2019) 9436 – 9448, https://doi.org/10.20964/2019.10.34

Taer, Erman, Friska Febriyanti, Widya Sinta Mustika, Rika Taslim, Agustino Agustino, and Apriwandi Apriwandi. Enhancing the performance of supercapacitor electrode from chemical activation of carbon nanofibers derived Areca catechu husk via one-stage integrated pyrolysis. Carbon Letters 31 (2021): 601–612.

Tavadi, Ashokkumar R., Dedapeer Basheer, Kassahun Gashu Melese, K. Kumaresan, N. Mohan, Mohammad Rafi H. Kerur, and L. C. Shashidhara. Experimental analysis of mechanical properties and FTIR analysis of areca fiber-reinforced epoxy composites incorporating Al_2O3. *Journal of Industrial Textiles* 55 (2025): 15280837241313217.

Thamba, Narendiranath Babu, Tanay Kuclourya, Mohit Kumar Jain, and Ramalinga Viswanathan Mangalaraja. Flexural properties of areca nut, sunn hemp and e-glass fibers reinforced with epoxy composites. *Materials Research Express* 6 (10) (2019): 105344.

Tharayil, Jeevan Mathew, and Prakash Chinnaiyan. Sustainable waste valorisation: Novel Areca catechu L. husk biochar for anthraquinone dye adsorption-Characterization, modelling, kinetics, and isotherm studies. Results in Engineering 20 (2023): 101624.

Thilagham, K. T., G. Gayathiri Devi, A. Kadirvel, and D. Kumar. Development of wheat husk biosilica and characterization of its areca reinforced polyester composite. *Biomass Conversion and Biorefinery* 14 (14) (2024): 15483–15492.

Tian, Zi-Han, Jueng-Tsueng Weng, Li-Jane Shih, An-Ci Siao, Tsai-Yun Chan, Yi-Wei Tsuei, Yow-Chii Kuo, Tsu-Shing Wang, and Yung-Hsi Kao. Arecoline inhibits the growth of 3T3-L1 preadipocytes via AMP-activated protein kinase and reactive oxygen species pathways. *PLoS One* 13 (7) (2018): e0200508.

Tiwari, Diwakar, Lalhmunsiama, and Seung-Mok Lee. Iron-impregnated activated carbons precursor to rice hulls and areca nut waste in the remediation of Cu (II) and Pb (II) contaminated waters: a physico-chemical studies. Desalination and Water Treatment 53 (6) (2015): 1591–1605.

Tong, Ting, Aiqing Xu, Shuhua Tan, Hengzhi Jiang, Lixin Liu, Senwen Deng, and Haihua Wang. Biological effects and biomedical applications of areca nut and its extract. *Pharmaceuticals* 17 (2) (2024): 228.

Tsakalova, Marinella, Cristina Onorato, Alexandros Zoikis Karathanasis, and Ioanna Deligkiozi. A circular economy framework for the assessment of bio-based value chains. *Sustainable Chemistry for the Environment* 6 (2024): 100099.

Tseng, Shuei-Kuen, Mei-Chi Chang, Ming-Lun Hsu, Cheng-Yao Su, Lin-Yang Chi, Wen-Chien Lan, and Jiiang-Huei Jeng. Arecoline inhibits endothelial cell growth and migration and the attachment to mononuclear cells. *Journal of Dental Sciences* 9 (3) (2014): 258–264.

Ujjinappa, Santhosh, and Lingadhalli Krishnamurthy Sreepathi. Evaluation of Physico-Mechanical-Combustion Characteristics of Fuel Briquettes Made from Blends of Areca Nut Husk, Simarouba Seed Shell and Black Liquor. International Journal of Renewable Energy Development 7 (2) (2018).

Vannarath, Adhirashree, and Arun Kumar Thalla. Effects of chemical pretreatments on material solubilization of Areca catechu L. husk: Digestion, biodegradability, and kinetic studies for biogas yield. Journal of Environmental Management 316 (2022): 115322.

Vardhan, Harsh, Soumya Sasmal, and Kaustubha Mohanty. Saccharification and structural changes in Areca catechu husk fiber. *Biofuels, Bioproducts and Biorefining* 18 (5) (2024): 1495-1510.

Vardhan, Harsh, Ram Balak Mahato, Soumya Sasmal, and Kaustubha Mohanty. "Production of xylose from pre-treated husk of areca nut. *Journal of Natural Fibers* 19 (1) (2022): 131–144.

Varghese, Lijo, and GC Mohan Kumar. The Epoxy-based composites with Size-Fractionated Waste Areca Sheath: An Experimental Investigation on the Macroscopic and Vibrational Properties. *Fracture and Structural Integrity* 19 (71) (2025): 49–66.

Venkatachalaiah, Madhushri ATTIHALLI, Raju NAMA Vasudeva Setty, and Jobish Johns. Effect of compatibilizer on the properties of areca-fiber reinforced polypropylene composites. *Journal of Natural Fibers* 19 (17) (2022): 15261–15275.

Venkateshappa, Srinivasa Chikkol, Basavaraju Bennehalli, Mownesh Gadde Kenchappa, and Raghu Patel Gowda Ranganagowda. Flexural behaviour of areca fibers composites. *BioResources* 5 (3) (2010): 1846–1858.

Venkateshappa, Srinivasa Chikkol, Suresh Yalaburgi Jayadevappa, and Prema Kumar Wooday Puttiah. Mechanical behavior of areca fiber reinforced epoxy composites. *Advances in Polymer Technology* 31 (4) (2012): 319–330.

Vinay, S. P., and N. Chandrasekhar. Facile green chemistry synthesis of Ag nanoparticles using areca catechu extracts for the antimicrobial activity and photocatalytic degradation of methylene blue dye. *Materials Today: Proceedings* 9 (2019): 499-505.

Vinu, V., and N. N. Binitha. Lithium silicate based catalysts prepared using arecanut husk ash for biodiesel production from used cooking oil. Materials Today: Proceedings 25 (2020): 241-245.

Wahyuti, Wahyuti, Abdul Karim, Rugaiyah Andi Arfah, Muhammad Zakir, Maming Maming, and Maswati Baharuddin. Effectiveness of HNO3 and NaOH Pretreatment on Lignin Degradation in Areca Leaf Sheath Fibre (Areca catechu L.) for Bioethanol Production. *Automotive Experiences* 8 (2) (2025): 227–241.

Wang, Chia-Chi, Pei-Ling Huang, Tsung-Yun Liu, and Tong-Rong Jan. Highly oligomeric procyanidins from areca nut induce lymphocyte apoptosis via the depletion of intracellular thiols. *Toxicology in Vitro* 23 (7) (2009): 1234-1241.

Wang, Dingfa, Luli Zhou, Wei Li, Hanlin Zhou, and Guanyu Hou. Anticoccidial effects of areca nut (Areca catechu L.) extract on broiler chicks experimentally infected with Eimeria tenella. *Experimental parasitology* 184 (2018): 16–21.

Wang, Ruimin, Feibing Pan, Ruiping He, Fengjiao Kuang, Lu Wang, and Xue Lin. Arecanut (Areca catechu L.) seed extracts extracted by conventional and eco-friendly solvents: Relation between phytochemical compositions and biological activities by multivariate analysis. *Journal of Applied Research on Medicinal and Aromatic Plants* 25 (2021): 100336.

Wang, Zihan, Ziyuan Guo, Yinghua Luo, Lingjun Ma, Xiaosong Hu, Fang Chen, and Daotong Li. A review of the traditional uses, pharmacology, and toxicology of areca nut. *Phytomedicine* (2024): 156005.

Warnakulasuriya, S., and T. H. H. Chen. Areca nut and oral cancer: evidence from studies conducted in humans. *Journal of Dental Research* 101 (10) (2022): 1139-1146.

Wijayarathna, Jayani, Sampath Induruwa, Deepika Priyadarshani, Samudrika Wijayapala, and Rumesh Samarawickrama. Aqueous Extraction and Dyeing Behavior of Areca Nut (Areca catechu) Natural Dye for Cotton Fabric. (2023).

Wubneh, Fisseha, Rotich K. Gideon, Dacheng Wu, and Babu Km. Extraction and Characterization of Fibers from Corn Husk. *Journal of Natural Fibers* 19 (16) (2022): 12862–12869.

Yadav, Gaurav, Nidhi Yadav, and Md Ahmaruzzaman. Microwave-assisted synthesis of biodiesel by a green carbon-based heterogeneous catalyst derived from areca nut husk by one-pot hydrothermal carbonization. Scientific Reports 12 (1) (2022): 21455.

Yallappa, S., M. Shivakumar, K. L. Nagashree, M. S. Dharmaprakash, Ajayan Vinu, and Gurumurthy Hegde. Electrochemical determination of nitrite using catalyst free mesoporous carbon nanoparticles from bio renewable areca nut seeds. *Journal of the Electrochemical Society* 165 (10) (2018): H614.

Yang, Bowen, Haiming Chen, Weijun Chen, Wenxue Chen, Qiuping Zhong, Ming Zhang, and Jianfei Pei. Edible Quality Analysis of Different Areca Nuts: Compositions, Texture Characteristics and Flavor Release Behaviors. *Foods* 12 (9) (2023): 1749.

Yong, F. S. J., S. H. Loh, H. M. Yusoff, A. Ahmad, P. W. Chia, K. Bakar, S. Mohamad, R. K. Liew, and S-Y. Kan. Areca catechu husk oriented activated carbon in application for Bisphenol A removal through adsorption method. *International Journal of Environmental Science and Technology* (2024): 1–10.

Yuan, Jianbo, Haonan Zhang, Hui Zhao, Hao Ren, and Huamin Zhai. Study on Dissociation and Chemical Structural Characteristics of Areca Nut Husk. Molecules 28 (3) (2023): 1513.

Yuan, Hanwen, Mengru Cao, Pan Yi, Qingling Xie, Yuqing Jian, Bin Li, Yan Qin et al. Determination of alkaloids and phenols in the chewable husk products of Areca catechu L. Using HPLC-UV and UHPLC-MS/MS. *Journal of Liquid Chromatography & Related Technologies* 41, no. 10 (2018): 612–620.

Yuan, Shyng-Shiou F., Leong-Perng Chan, Hieu DH Nguyen, Chang-Wei Su, Yuk-Kwan Chen, Jeff Yi-Fu Chen, Shigetaka Shimodaira, Stephen Chu-Sung Hu, Steven Lo, and Yen-Yun Wang. Areca nut-induced metabolic reprogramming and M2 differentiation promote OPMD malignant transformation. *Journal of Experimental & Clinical Cancer Research* 43 (1) (2024): 233.

Yusriah, L., and S. M. Sapuan. Properties of betel nut husk reinforced vinyl ester composites. In *Natural Fibre Reinforced Vinyl Ester and Vinyl Polymer Composites*, pp. 129–155. Woodhead Publishing, 2018.

Zhao, Hongliang, Jinghao Yang, Jun Zeng, Bosheng Zhou, Min Yang, Xiaohong Yang, and Ranfeng Sun. Development of subcritical water extraction for areca alkaloids and its influence on the structure of areca nut husk." *Molecules* 30, no. 4 (2025): 886.

Zhao, Xin, Ruo-Chen Li, Wen-Xuan Liu, Wen-Sheng Liu, Ying-Hao Xue, Ren-Hua Sun, Yu-Xin Wei et al. Estimation of crop residue production and its contribution to carbon neutrality in China. *Resources, Conservation and Recycling* 203 (2024): 107450.

Zheng, Wei, Xiao-ming Li, Fei Wang, Qi Yang, Pin Deng, and Guang-ming Zeng. Adsorption removal of cadmium and copper from aqueous solution by areca—a food waste. Journal of Hazardous Materials 157 (2–3) (2008): 490–495.

https://ajpp.in/uploaded/p358.pdf

The manufacturer's authorised representative in the EU is Springer Nature Customer Service Centre GmbH, Europaplatz 3, 69115 Heidelberg, Germany. If you have any concerns regarding our products, please contact ProductSafety@springernature.com

Printed and bound by CPI Group (UK) Ltd, Croydon, CR0 4YY
05/10/2026
02228035-0002

The manufacturer's authorised representative in the EU is Springer Nature Customer Service Centre GmbH, Europaplatz 3, 69115 Heidelberg, Germany. If you have any concerns regarding our products, please contact ProductSafety@springernature.com

Printed and bound by CPI Group (UK) Ltd, Croydon, CR0 4YY

05/10/2026

02228035-0002